3RD EUROPEAN SYMPOSIUM ON ENGINEERING CERAMICS

Proceedings of the 3rd European Symposium on Engineering Ceramics, held at the Regent Crest Hotel, London, 28–29 November 1989, organised by IBC Technical Services Ltd.

3RD EUROPEAN SYMPOSIUM ON ENGINEERING CERAMICS

Edited by

F. L. RILEY

Division of Ceramics, School of Materials, University of Leeds, UK

ELSEVIER APPLIED SCIENCE

LONDON and NEW YORK

ELSEVIER APPLIED SCIENCE PUBLISHERS LTD
Crown House, Linton Road, Barking, Essex IG11 8JU, England

Sole Distributor in the USA and Canada
ELSEVIER SCIENCE PUBLISHING CO., INC.
655 Avenue of the Americas, New York, NY 10010, USA

WITH 23 TABLES AND 123 ILLUSTRATIONS

SOFTCOVER REPRINT OF THE HARDCOVER 1ST EDITION 1991

British Library Cataloguing in Publication Data

European Symposium on Engineering Ceramics (3rd; 1989;
London, England)
3rd European Symposium on Engineering Ceramics.
1. Materials: Ceramics. Engineering aspects
I. Title II. Riley, F. L.
620.14

ISBN 978-94-011-7992-8 ISBN 978-94-011-7990-4 (eBook)
DOI 10.1007/978-94-011-7990-4

Library of Congress Cataloging-in-Publication Data

European Symposium on Engineering Ceramics (3rd: 1989:
London, England
3rd European symposium on engineering ceramics/edited by
F. L. Riley
p. cm.
Proceedings of the 3rd European Symposium on Engineering Ceramics, held at the Regent Crest Hotel, London, 28–29 November 1989. Organised by IBC Technical Services Ltd.
Includes bibliographical references and index.
ISBN 978-94-011-7992-8
1. Ceramics—Congresses. I. Riley, F. L. II. IBC Technical Services Ltd. III. Title.
TP786.E97 1989
620.1′4—dc20 90-46312
CIP

Preface

This volume is the proceedings of the 3rd European Symposium on Engineering Ceramics, held in London, 28–29 November 1989, under the auspices of IBC Technical Services Ltd. The Symposium sessions were chaired by Eric Briscoe, who also introduced the Symposium with the very appropriate review 'Ceramics in Europe'.

The term 'engineering ceramics' is commonly taken to mean a group of special high-strength and heat-resistant ceramic materials developed almost exclusively for the advanced internal combustion engine of the next century. It is not always fully appreciated that high grade fine microstructure ceramics both of the oxide and of the non-oxide classes, whether they be termed engineering, fine, special, advanced, structural or technical, have been supporting a large number of diverse and profitable industries over many decades. Indeed, in some respects these materials can be regarded as natural developments from the long-established refractories field, and the distinction between an engineering ceramic and a refractory can become blurred, as the contribution in this volume on 'Nitride Bonded Carbide Engineered Ceramics' shows. It is of significance that in Japan, for example, much development work in the engineering ceramics field was initiated by the refractories industries, seeking to diversify possibly but doing so on the basis of long experience in the refractories area.

The main objective of this Symposium was to help engineers and designers to assess the present state of the field of engineering ceramics. The programme was intended to be low-key, factual, balanced and critical. To achieve this objective a team of speakers from the engineering ceramics industries and closely associated laboratories, each member a widely respected authority in his own area, was invited to present accounts of specific materials and applications, and related development work. The majority of the twelve invited contributions to the programme placed emphasis on current applications, but in doing so drew attention to recent background developments in order to provide guidance to the avenues likely to be of importance for the near-term future. It was not the intention to concentrate on the more spectacular applications in the engine component area, though these were not ignored. Nor was it intended that points of scientific detail would be examined in depth; there are many other occasions in the scientific meeting calendar providing ample opportunity for this approach. This programme, it was hoped, would provide participants with broad over-views interwoven with illustrations of specific applications and materials' development programmes.

The two-day Symposium took a selected group of subjects and materials for a critical review on the basis of a broad European perspective. It is hoped that the selection presented in this volume will provide a balanced guide to the subject of engineering ceramics. Oxides and non-oxides are reviewed, as are the newer composites, and the now maturing transformation toughened zirconias. There are reflective contributions on how successful production processes can be developed, as well as forward-looking overviews of new processing methods. Established applications of ceramics in the important areas of wear and abrasion resistant materials are also reviewed, taking in established uses in powder and slurry handling, and the newer ceramic bearing area. Very broad reviews of developments in supporting funding for work on the engineering ceramics in Europe, and on recent developments in 'fine' ceramics in Japan were presented by Eric Briscoe and by Keiji Matsuhiro respectively.

The assumption is made that the reader of this proceedings volume is scientifically educated, but non-specialist, and wishes to bring him or herself up-to-date on selected aspects of the subject of engineering ceramics through a structured series of authoritative reviews. This description of a reader can be applied equally to the materials technologist, or company director, and the undergraduate or post-graduate student whose programme includes treatment of the structural or engineering ceramics. Many of the contributions are extensively referenced, and serve as a guide to wider reading.

F. L. RILEY

Contents

List of Contributors

I. BIRKBY
Dynamic–Ceramic Ltd, Bournes Bank, Burslem, Stoke-on-Trent, Staffordshire, ST6 3DW, UK

E. M. BRISCOE
Watersmeet, Fradley Junction, Alrewas, Burton-on-Trent, Staffordshire, DE13 7DN, UK

F. CAMBIER
Centre de Recherches de l'Industrie Belge de la Céramique (CRIBC), 4 Avenue Gouverneur Cornez, 7000 Mons, Belgium

P. DESCAMPS
Centre de Recherches de l'Industrie Belge de la Céramique (CRIBC), 4 Avenue Gouverneur Cornez, 7000 Mons, Belgium

C. R. DIMOND
Morgan Roctec Ltd, Bewdley Road, Stourport-on-Severn, Worcestershire, DY13 8QR, UK

P. FEINLE
Elektroschmelzwerk Kempten GmbH, Postfach 1526, 8960 Kempton, Federal Republic of Germany

R. GANZ
Hoechst AG, FTT Neue Technologien, Postfach 80 03 20, 6230 Frankfurt/Main, Federal Republic of Germany

J. HEINRICH
Hoechst CeramTec AG, Werk Selb, Wilhemstr. 14, 8672 Selb, Federal Republic of Germany

O. HEINZ
Hoechst AG, FTT Neue Technologien, Postfach 80 03 20, 6230 Frankfurt/Main, Federal Republic of Germany

H. HODGSON
Dynamic–Ceramic Ltd, Bournes Bank, Burslem, Stoke-on-Trent, Staffordshire, ST6 3DW, UK

G. E. HOLLING
Thor Ceramics Ltd, PO Box 3, Stanford Street, Clydebank, G81 1RW, UK

S. A. HORTON
SKF Engineering and Research Centre B.V., Postbus 2350, 3430 DT Nieuwegein, The Netherlands

J. HUBER
Hoechst CeramTec AG, Werk Selb, Wilhemstr. 14, 8672 Selb, Federal Republic of Germany

K. KENDALL
ICI, PO Box 11, Runcorn, Cheshire, WA7 4QE, UK

H. KNOCH
Elektroschmelzwerk, Kempton GmbH, Postfach 1526, 8960 Kempton, Federal Republic of Germany

R. J. LUMBY
103 Solihull Road, Shirley, West Midlands, B90 3HW, UK

K. MATSUHIRO
NGK Insulators Ltd, Nagoya, Japan. Present address: NGK Europe GmbH, Morgenthalerallee 77–81, D-6236 Eschborn/Ts., Federal Republic of Germany

M. MATSUI
NGK Insulators Ltd, Nagoya, Japan

G. C. PADGETT
British Ceramic Research Ltd, Queens Road, Penkhull, Stoke-on-Trent, ST4 7LQ, UK. Present address: *'Hunters Tryste', Tower Road, Ashley Heath, Market Drayton, Shropshire TF9 4PU, UK*

H. SCHELTER
Hoechst CeramTec AG, Werk Selb, Wilhemstr. 14, 8672 Selb, Federal Republic of Germany

J. TIRLOCQ
Centre de Recherches de l'Industrie Belge de la Céramique (CRIBC), 4 Avenue Gouverneur Cornez, 7000 Mons, Belgium

1

Ceramics in Europe — An Overview

E. M. BRISCOE

Watersmeet, Fradley Junction, Alrewas, Burton-on-Trent, Staffordshire, DE13 7DN, UK

1. INTRODUCTION

This is the third conference of this title spanning some 5 years. In giving the Opening Address I will, no doubt, raise as many questions as explanations for the apparently low level of additional exploitation relative to the considerable input in terms of research and development (R and D) expenditure. It is, however, not uncommon for many decades to pass before new or improved materials find their way into common engineering use. The transistor was originally conceived in the 1930s and it required the impetus of national defence crises worldwide to push its development into sophisticated products; development costs were 'lost'. Development of the transistor into cost-effective civil products for wider use was therefore given a 'free investment'.

Today there are other pressures to achieve unique solutions to what have now become engineering problems, as a result of intense international industrial competition and the consumer demand arising from the much higher standard of living available to a larger part of the world. Significantly, governments worldwide recognise that intervention on a national scale is necessary to support innovation in materials on account of the huge costs and uncertainties. Some 'bean counters', if I may use a term understood by those of us who are not accountants, will naturally have more difficulty than others in understanding the commercial wisdom of expenditure for which no short-term high return certainty is assured. Consequently funding or long term and speculative R and D in materials development is not likely to be carried out quickly enough, or indeed at all, unless the financial burdens and risks are shared by the taxpayer. Even where science and technology is admired, government pays to ensure modern and competitive industry to support its citizens in the long term.

It seems that a crisis of one kind or another is needed to catalyse action. The dramatic rise in oil prices in the 1970s was undoubtedly such a catalyst for the industrial nations to develop economies in the use of oil-related consumption devices; governments worldwide co-ordinated such initiatives and provided significant financial support.

Prior to the oil price crisis, there was another catalyst, of an environmental nature, for an engineering solution requiring enhanced properties of materials; the impending 1970 Clean Air Act in the USA. In this case there was an acceleration of work on ceramics as a potential solution to environmental needs, to reduce the vehicle exhaust generated 'smog', of Los Angeles in particular. It has taken a long time for Europe to get to grips with the same fundamental problem, but the need to do so is now with us.

As long ago as the early 1950s, the then UK Ministry of Supply funded the development of silicon nitride for 'undisclosed' reasons. Subsequently in 1956 the UK Admiralty Materials Laboratory at Holton Heath commenced the development of a silicon nitride internal combustion engine requiring no liquid cooling. The size of the investment made is not revealed, but in terms of R and D investment on the new high temperature and strong ceramics it was probably significant in world terms at that time.

Other initiatives, spurred on by quadrupled oil prices, followed rapidly and were aimed at significantly improving the thermal efficiency of the internal combustion engine. The development of the 'adiabatic', or minimum heat loss, engine became the worldwide target, and this required materials with properties not possessed by metals.

2. INTERNATIONAL PROGRAMMES AND GOVERNMENT FINANCIAL SUPPORT

A summary of the programmes in the UK and USA is given in Figs 1 and 2.

Some idea of the size of investment in the USA is shown in Fig. 3 and that of the UK and EEC in Fig. 4, which takes us through the period up to today, apart from developments which are more or less classified or

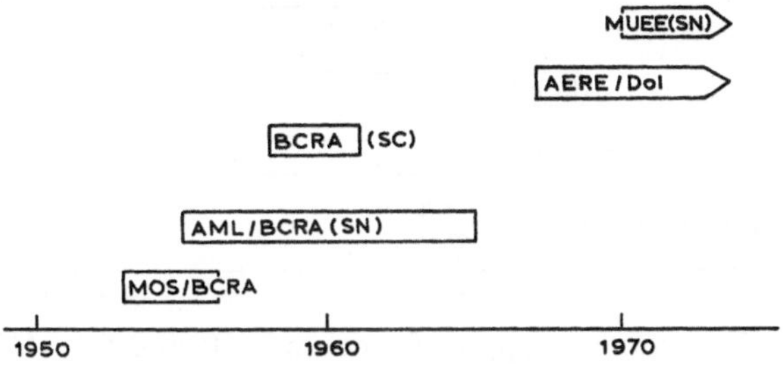

Fig. 1. Government support for ceramics R and D — UK.

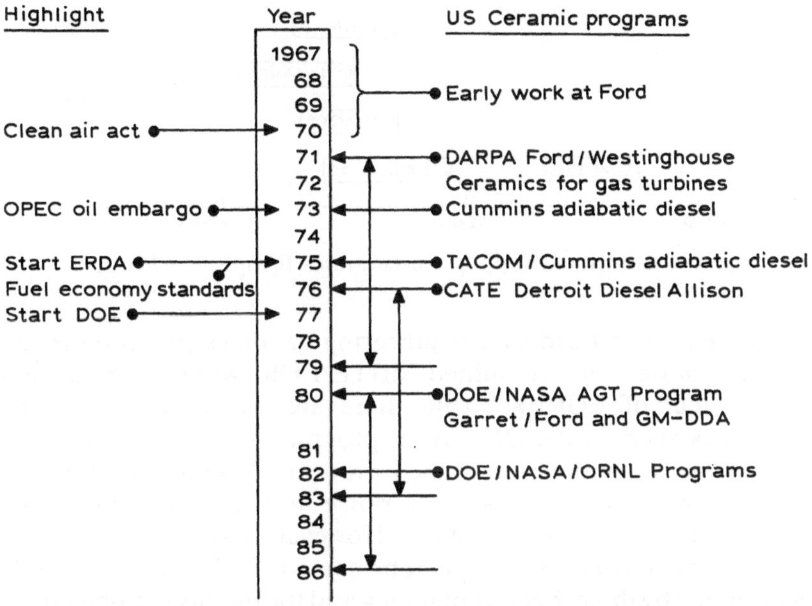

Fig. 2. Highlights and related US ceramic programmes.

commercially sensitive. The acronyms describing the projects were no doubt designed to confuse the 'competition' and with the passing of time some of us, though familiar with the derivations at the time, are now also confused with regard to the very obvious! (One has to forgive some lightheartedness, as a sense of humour is an advantage when tackling what were then 'impossible' tasks, some of which remain as intractable today.)

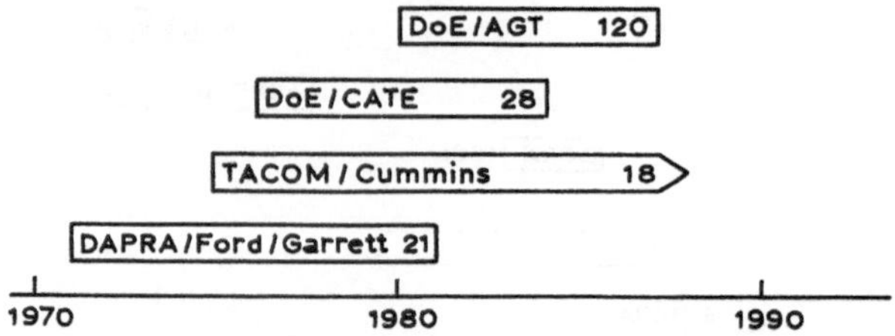

Fig. 3. Government support for ceramics R and D — USA.

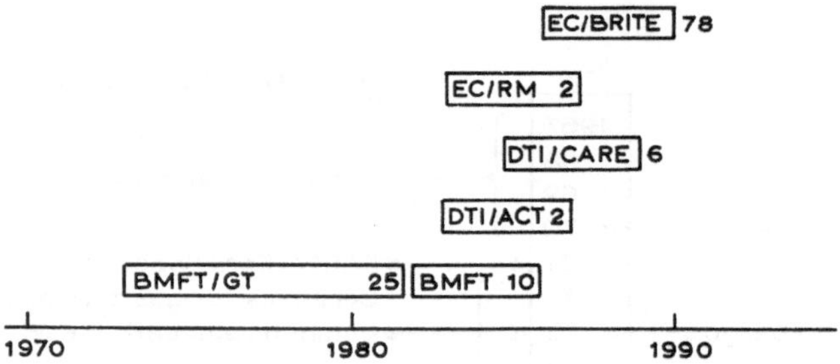

Fig. 4. Government support for ceramics R and D — Europe.

Today the latest European engineering ceramics programmes are principally those of the combined BRITE/EURAM programmes, and indirectly that of EUREKA (but there are many purely national programmes such as the UK CARE programme which finishes this year). The latest Commission of the European Communities initiative extends across the whole materials range and the split between the essential components is not settled. However, it is possible that the ceramics content over 3 years might be some £17 M, if one divides the sum available by three. Factors of policy and the quality of applications also have to be taken into account of course. That sums of this size are considered to be strategically necessary for the competitiveness of Europe can be taken as implying that there are many solutions still required and that it is important to achieve the remaining important objectives. This Symposium will address itself to some of those issues. However, let us not assume that the further development of engineering ceramics is a task for Europe alone. Japan has had a series of well-conceived and managed programmes under the aegis of MITI funding and as we may hear from a distinguished speaker from NGK, Japan currently has substantial funding of a ceramic component gas turbine

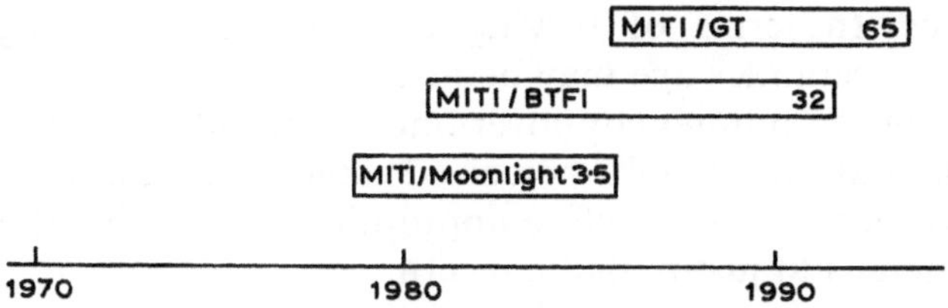

Fig. 5. Government support for ceramics R and D — Japan.

of automotive size. Some of us will recall the Rover gas turbine which ran at Le Mans in the early 1970s. Sadly this may be yet another example of a British innovation killed off by short term policies. Figure 5 relates to Japan.

What then is the scenario that directs the taxpayer's money into the development of materials? We do not have to look further than back at history — the Ages of Man are defined traditionally in terms of materials, from the Stone Age to today, see Fig. 6. On the assumption that there is no better prediction than that arising from hindsight, it is not unreasonable to predict that when looking back at the period 1950–2000 the most influential material classifications will have been ceramics

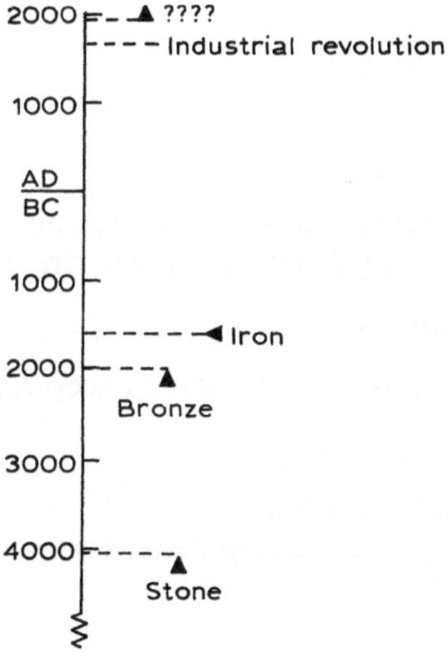

Fig. 6. Materials 'ages'.

and plastics. In terms of solving the most difficult engineering problems, the ceramics age may be just ahead.

The level of commitment by governments to fund Ceramics R and D is in itself an indication of the perceived value to the respective nations in the longer term. It is therefore appropriate to examine the techno/economic details to justify that concern.

Our expert speakers today will present their own version of the case, some with a more commercial emphasis, and others from a purely scientific viewpoint. It is by presenting that essential balance that commercial progress is made. After all, it is by the process of exposing apparently uneconomic scientific discovery to the harsh economic environment of life that real progress is made; one without the other would have left us in the Stone Age. Though the experimenters with a piece of flint producing a spark could not have understood the scientific basis of their discovery, to them it was an important means of survival. Perhaps the same applies today?

3. THE TECHNO/ECONOMIC CASE FOR ENGINEERING CERAMICS

3.1 General Properties: Ceramics Compared with Metals

High values for ceramics:

compressive strength
hardness
stiffness
corrosion resistance
retained high temperature tensile strength
electrical resistivity (some, however, are good conductors).

Low values for ceramics:

thermal conductivity (with some exceptions)
specific heat
coefficient of expansion
density (some exceptions)
toughness (but some getting better for example, some zirconia based materials)
tensile strength.

(But see Section 4 concerning 'specific' properties, for the effect of density.)

3.2 Cost

Competitive cost for simple shapes where unique properties are vital. High cost for complex engineering accuracy for shapes relative to the same metal shape. Re-design for material advantages can change the relative cost. There may be no need to apply protective coats, for example.

3.3 Raw Material Availability

The chemical constituents of materials are generally readily available. Some special components may not be available. Processing costs can, however, be high, but the scale of demand is likely to reduce these.

4. SPECIFIC PROPERTIES

It is not intended to list a comprehensive data bank. For the purpose of this paper it is more useful to make comparisons with other, more commonly used, typical engineering materials. My 5-year old zirconia hammer which knocks 15 cm nails into wood always surprises engineers who look for an explanation other than the simple truth!

Typical properties with which the mechanical engineer will normally be concerned include tensile strength, stiffness, and to an increasing extent, the minimum mass required to satisfy design criteria, that is related to density. To make the comparisons on the above basis, specific properties can be calculated, that is property value divided by density. Figure 7 shows the effect of such calculations on the ranking of some materials.

From the above the following conclusions can be drawn:

a high strength polymer composite is strongest (but has a relatively low maximum operating temperature);
silicon nitride is nearly as strong as mild steel (and can operate at much higher temperature and in composite form is stronger, and tougher).

Other similar calculations show that the flexibility of transformation toughened zirconia approaches that of steel, for example.

Toughness is inevitably the question engineers will raise. Ceramics have a fundamental problem with this property, but some forms of zirconia are as tough as the brittle cast iron and engineers long ago learned how to use this material. Ceramic composites have been made

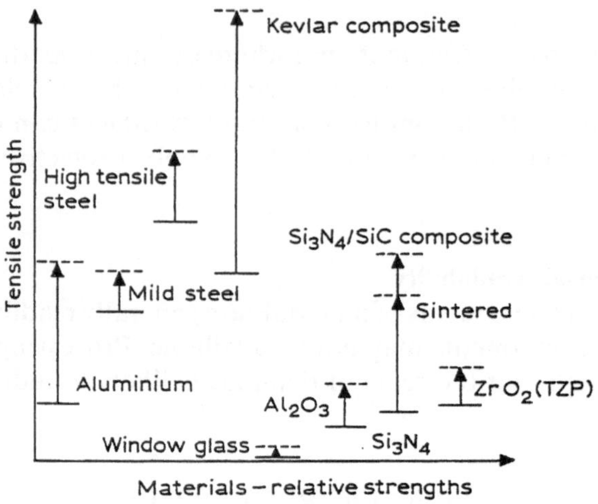

Fig. 7. The relative (——) absolute and (---) specific strengths of some materials.

of toughness higher than ordinary steel though with great difficulty, and at high cost. They are, however, less dense.

Our speakers today will provide specific information on many of these points.

Tensile strength at high temperature is becoming even more important. There is an increasing number of applications where an increase in operating temperature is necessary or advantageous, for example in chemical processing, and in heat engines, in order to raise process performance or thermal efficiency. Figure 8 shows the typical tensile behaviour of some ceramics and metals with increase in temperature. There are virtually no metal alloys capable of withstanding useful tensile loading beyond 1250°C. In fact creep restricts the temperature of the best (nimonics) to ~1000°C. Certain ceramics are capable of at least that level and retain other useful attributes not possessed by metals. However, the toughness of ceramics is still below that of competitor materials. The catastrophic mode of failure of ceramics is also the real concern, and therefore Weibull moduli of the order of 30 are required before engineers will be prepared to consider the use of ceramics in high temperature critical designs. Ceramic/ceramic matrix composites are therefore an important area for research as they probably are the solution to high toughness and strength at temperatures beyond those at which metals can be used: hence the

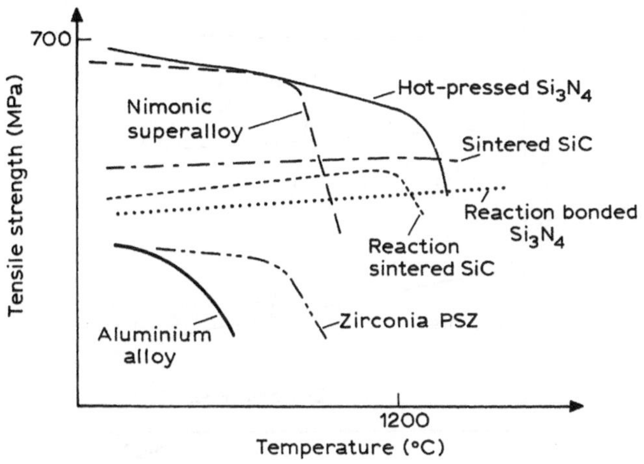

Fig. 8. Typical tensile strength as a function of temperature.

interest in certain components in gas turbines. Figure 9 suggests the relative 'places' of metals and ceramics in general terms of temperature/stress.

Low thermal conductivity and coefficient of expansion are rarely matched by metals. There are exceptions. A high thermal conductivity coupled with other advantages of ceramics, as found in silicon carbide, can also make it a unique material for electronic applications and heat exchange devices. Aluminium titanate has one of the lowest conductivities and coefficients of expansion, but has very low strength. However, by putting it into compression by casting metal around it, for example, it can be a valuable high temperature insulating material, especially on account of its good resistance to thermal shock.

5. LOGICAL TECHNIQUE FOR APPLICATION ASSESSMENT

By constructing a matrix of ceramic properties and engineering parameter requirements one can assess whether a ceramic is likely to be satisfactory in a specified application; subsequent reference to specific properties will confirm or rule out that possibility. It is obviously necessary that the desired properties reside in the same ceramic. For instance, zirconia is close to cast iron in coefficient of thermal expansion, and it is a very good thermal insulator. It is therefore a good

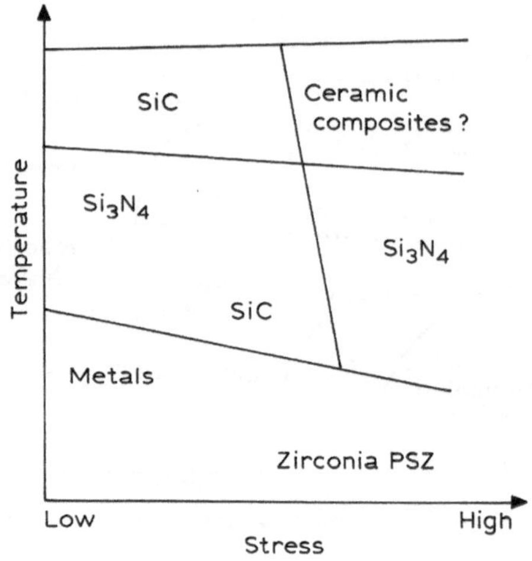

Fig. 9. The relative 'places' of metals and ceramics in general terms of temperature and stress.

candidate for attachment to metals at elevated temperature. However its density is high, and if the component is subject also to high acceleration, then zirconia may not be suitable. Otherwise zirconia might have been a good choice of material for high temperature combustion reciprocating components. Furthermore the maximum operating temperature of use for a transformation toughened zirconia of about 800°C though not high for a ceramic, is high enough for most engineering purposes. A rather more sophisticated matrix is therefore required where property levels are needed in the same material. This can be set up using a simple computer programme into which a data bank is fed along with the ranges of engineering design parameters. Then decisions can be taken rapidly and reliably. Experience in various parts of the world has shown, however, that in spite of the sophisticated computerised design facilities now available, coupled with equally sophisticated property measurement equipment, there is no substitute finally for a demonstration device.

Nature still has some surprises in store. Those who have worked on the 'adiabatic' engine will know that the thermal efficiency gains, though useful, are less than expectation, for unexpected reasons.

It will thus be seen that there are certain properties which require improvement:

reliability (less tendency to sudden failure);
higher tensile strength at low and moderate temperatures;
higher toughness;
easier production of complex shapes to close tolerances and therefore of lower cost.

The current CEC BRITE/EURAM programme recognises the above fundamental needs, and some details of the themes to be funded as given in the next section.

6. THE BRITE/EURAM PROGRAMME 1990/1992

The preceeding section attempts to justify the need for government intervention to encourage, co-ordinate and support financially the difficult task of developing the advanced engineering ceramics. The technical content bias is that seen through the eyes of the engineer responsible for initiation and profitable management of a medium-size manufacturing industry, employing advanced ceramics both in their manufacture and in their incorporation with metals and plastics to form 'total' end-products. There may be other views, but I suspect that as much has already been spent on the necessary fundamentals, we shall be forced to focus on the science needed to support identified end-product requirements, that is that developments will require to be more end-product led. It is now up to the engineer, trained predominantly to use metals, to learn how to use less tolerant, but otherwise more attractive, properties, and to the scientist to produce more 'metal-like' properties into ceramics. This is easier said than done!

The need for greater tensile strength, toughness coupled with greater reliability, less catastrophic failure, and cost effective manufacture, is a view shared by many engineers faced with more advanced requirements, either in terms of product cost or performance, especially at temperatures rising towards those at which metals fail. It is also in this area that relatively high added-value will arise. The other prospect, evident in the foregoing, is that of capitalising on the hardness and corrosion resistance properties of ceramics; tribological applications already exist and there must be more to come, using coatings and wear resistant surfaces, for example. Applications using such features are now established, and there is growth potential in this area.

The themes of the current BRITE/EURAM programme are

necessarily wider than the personal focusing in the technical views expressed above. Collaborative research will be supported in the following areas:

advanced materials technologies,
design methodology and assurance of products and processes,
applications of manufacturing systems,
technologies for manufacturing processes.

There are two main types of action:

industrial applied research (industrial organisations-led),
focused fundamental research (University-led with industrial approval).

There are requirements for cross-border collaboration, and of minimum size of project for example. The official information package gives full details. There may be some revision to the launch document when the call is announced for the next tranche of proposals — probably in May 1990, with closing date for submissions in September 1990.

It is probable that more emphasis will be placed on high potential, economic impact, industrial relevance. By way of giving general advice the following considerations could be taken into account.

In spite of very clear rules stated in the CEC document, it is surprising how often applicants ignore some of the requirements. The 'success rate' may only be of the order of 15%, not necessarily because of 'rationing' by the Commission, but because the applications fail on the rules or in terms of quality. The official language is English, and sometimes our Continental brothers/sisters write better and clearer English! The Assessors who 'mark' the applications are from a wide range of member states. In my experience, as an Assessor, there is normally only one for each subject group for assessment, whose native tongue is English. Badly written English is more difficult for a non-UK citizen to understand as he/she will have learned 'good' English. On the other hand, a non-UK project leader whose English is not good can let down an otherwise very good application. In that event the UK collaborator could tactfully and with advantage, do the final editing.

'Presentation', that is the overall impression that the application gives, with regard to layout, diagrams, references, ease of finding the salient points for example, is undoubtedly of some importance. The assessment team may have to deal with an average of eight applications in the course of a working day (sometimes 10 h long!), and over 10 days non-stop. Some well-chosen phrases go a long way to help to crystallise one's impression of a sometimes long and complex document. Make it easy for the Assessors to recognise a good case! Make no mistake, it is the Assessors who decide the fate of the applications, without national favour, because the names and nationalities of the applicants are not known until an irrevocable marking has taken place.

For those concerned about confidentiality, each Assessor signs a non-disclosure agreement, and he/she works for the Commission and not his/her own State. In my experience, the Assessors are quite non-partisan and it sometimes occurs that we frequently find that we are the fiercest critics of our own Country's applications. So expect no favours!

Finally, I surmise that since each nation is putting in substantial contributions into the BRITE/EURAM programme, there will be less on offer from their own national Exchequer. So each member state will increasingly find that Brussels is the source of funding for R and D.

In conclusion, the current BRITE/EURAM programme may have some £17 million to award over the next 2 years on a 50% funding basis on the supposition that ceramics takes 33% of the remaining funds; a substantial amount not likely to be found elsewhere.

I recommend therefore that the Commissioners' announcements in due course be read carefully. There is obviously an attraction to the 'bean counters' to only have to fund 50% of your ambitions. Meanwhile the initiation or further development of rapport with colleagues across the water will not come amiss as the time between the 'call' and 'submission' is not long even if you have a 'warm' start and very short if you have not yet 'crossed the Channel', either way!

ACKNOWLEDGEMENTS

The author is indebted to the following: Dr Tom Sinclair, The Department of Trade and Industry, London, for the data contained in Figs 1–5. Mr G. Meetham, Chief of Non-Metal and Composite Manufacturing Technology, Rolls Royce, Derby, for some data contained in Fig. 8. Dr Hiroshi Abe, Asahi Glass, Japan, for some data contained in Fig. 9 and which was originally presented at the 37th Annual Meeting of the American Ceramic Society on October 29th 1984.

2

Review of the National and International Standardization of Engineering Ceramics

GERALD C. PADGETT

British Ceramic Research Ltd, Queens Road, Penkhull, Stoke-on-Trent, ST4 7LQ, UK

1. INTRODUCTION

Standardization is a means whereby materials, products and systems are described and evaluated in a standard or specified manner. Standards themselves, whether national or international, must conform to two working rules. They must firstly not inhibit progress and the development of new products and secondly, they should not prevent healthy competition. This latter point emphasizes the possibility of using National Standards to protect home markets and to frustrate imports.

Standardization itself is a relatively new concept and can be considered as one of the consequences of the Industrial Revolution. It started around the beginning of this century with almost simultaneous activity in the USA and Europe. The leader in the race was the American Society for Testing and Materials (ASTM), which was founded in 1898. ASTM's first standard, called 'Standard Specification for Steel Rails', was published in May 1900. Within Europe, the first official standards body was in the UK, where in 1901 the original Engineering Standards Committee was formed as the forerunner of the British Standards Institution (BSI), its first terms of reference being the standardization of steel sections. In Germany, the Standards Committee of German Industry, (now the Deutsches Institüt für Normung — DIN), was founded in May 1917. In France, the creation of the Permanent Standardization Committee in 1918 preceded the foundation of the Association Francaise de Normalisation, (AFNOR), in 1926. It is perhaps no coincidence that the French also started work first on the standardization of rails and steel sections.

In Japan, industrial standardization started rather late compared with both Europe and the USA. The concept started with Government standards for the procurement of commodities by the army and navy during the Meiji Era (1868–1912), and standard specifications for steel water pipes at the beginning of the Taisho Era (1912–1926). The Japanese Engineering Standards Committee (predecessor of JISC), was founded in 1921 and this committee established 520 Japanese Engineering Standards (JES), by April 1941. During the Second World War, the committee set up Temporary JES with simplified contents and procedures. Also during the war, 660 Japanese Aircraft Standards were established under the Aircraft Manufacturing Service Law. After the war, the existing JES were studied and revised and a start was made to establish new standards. This culminated in the formation of the Japanese Industrial Standards Committee (JISC), in 1949. This committee deliberates on JIS and is controlled and supported by the Government Agency of Industrial Science and Technology (AIST). AIST forms part of the Government Ministry of International Trade and Industry (MITI).

The start of international standardization was brought about by the electrical industry, a comparative newcomer to the industrial scene. That industry provided the first real step in the promotion of international standardization as we now know it, in the founding between 1904 and 1906 of the International Electrotechnical Commission (IEC). In April 1926, the 18 countries which had then set up national standards bodies met together to consider the extension of international collaboration to other fields. This resulted in the formation of the International Federation of the National Standardising Associations (ISA), with most of its strength coming from Europe. ISA was not particularly effective and it ceased to exist when war broke out in 1939. After the war, a new standards co-ordinating committee was set up under the United Nations banner. A full scale conference was held in London in October 1946, at which representatives of 25 standards bodies agreed to set up the International Organisation for Standardization (ISO). The IEC became its electrical division, though having full technical and financial autonomy. ISO has proved to be a success and now has members comprising the national standards organizations of 90 countries.

European standardization was initiated in 1960 when a meeting of the West European standards bodies agreed that the then Six and Seven should actively collaborate in unification of their standards within the wider ISO framework. The outcome of this decision was the creation of CEN (European Standards Coordinating Committee), and CENELEC, its counterpart in the electrotechnical field. Co-operation agreements between CEN/CENELEC and the Commission of European Communities drawn up in 1984 enabled the commission to give mandates to the joint European Standards Institute for specific projects.

2. STANDARDIZATION OF ENGINEERING CERAMICS

Certain difficulties arise due to terminology as to what is meant by 'advanced' ceramics. For the purpose of this review it will refer to those ceramics which are suitable for high performance application. There already exist applications in the electrotechnical field which are already covered by standards. These exist in national standards and also within IEC.

The first country to standardize these ceramics as materials in their own right has been Japan. Beginning in 1981, they have published five standards and these are listed below:

JIS R 1601 — 1981	Testing Method for Flexural Strength (Modulus of Rupture) of High Performance Ceramics.
JIS R 1602 — 1986	Testing Methods for Elastic Modulus of High Performance Ceramics.
JIS R 1603 — 1987	Methods for Chemical Analysis of Silicon Nitride Powders.
JIS R 1604 — 1987	Testing Method for Flexural Strength (Modulus of Rupture) of High Performance Ceramics at Elevated Temperature.
JIS R 1605 — 1988	Method for Elastic Modulus of High Performance Ceramics at Elevated Temperature.

The JIS work is concentrating on the standardization of test methods and has an on-going programme of 15 separate items which is supervised by the Japan Fine Ceramics Association (JFCA).

The BSI formed its 'engineering' ceramics committee in 1985. Its initial terms of reference were to develop standard methods of test and currently has seven standards which are close to publication. Current work now included in the BSI work programme is the preparation of a glossary of terms. The creation of the BSI committee was very closely followed by the setting up of ASTM Committee C-28 on Advanced Ceramics in the USA. This committee has a comprehensive work programme which is broken down into the following areas: Performance, Properties, Processing, Design & Evaluation and Characterization. The ASTM will shortly issue a standard 'Test Method for Flexural Strength of Advanced Ceramics at Ambient Temperature', which is essentially an up-date of an earlier Military Standard MIL-STD-1942 (MR). Other work which is close to publication is a comprehensive document on nomenclature. Other national standards organizations which are known to be active with regard to advanced ceramics include AFNOR (France) and DIN (FRG). A summary of on-going activities, prepared by a European Ad-Hoc Committee, is given in Table 1.

At present there is no official standardization of advanced ceramics under the framework of the ISO. There is, however, certain international work in VAMAS and also in a separate tripartite USA–Sweden–FRG agreement. VAMAS is the Versailles Project on Advanced Materials and Standards, the member countries being Canada, France, FRG, Italy, Japan, UK, USA and CEC. The work on ceramics is in Technical Working Area 3 under the chairmanship of Professor P. Boch of France.

There has been much interest in the standardization of advanced ceramics within Europe and within the CEC in particular. This has culminated in a mandated request from the Commission to both CEN and CENELEC for the establishment of a comprehensive programme for setting up European prestandards (ENV) and European Standards (EN) in the field of advanced industrial ceramics. It further requests that the programme shall be elaborated taking into account international standardization activities.

The Technical Board of CEN has acted very quickly and has agreed to create a new Technical Committee CEN/TC 184 with the provisional title of 'High Performance Ceramics'. Its proposed scope is: standardization in the field of high performance ceramics with specific initial tasks being terminology and classification; sampling; methods of test (including physical, chemical, mechanical, thermal, textural, electrical, etc.) for monolithic ceramics, ceramic composites, ceramic coatings, ceramic powders and ceramic fibres. As a material-related TC, it is expected that the TC will liaise closely with function-related TCs and standardizing bodies working in such fields as electrotechnology, biomedical (products), nuclear, optical, aerospace, the construction, etc. The secretariat of the new TC has been allocated to the BSI. In view of the urgency to provide a work programme for the CEC by the end of the year, the BSI are now preparing to host the inaugural meeting of TC 184 in September 1989 at the Manchester office of the BSI.

CENELEC have also been responsive to the mandated request from the Commission. They have called a meeting of experts from the various national bodies to determine whether a possible separate CENELEC work programme is required in the field of Advanced Ceramics. It is anticipated that potential work might belong to the fields of interest of a number of technical committees of the IEC.

3. THE PRESENT POSITION REGARDING ADVANCED CERAMICS STANDARDIZATION

The development of standards tends to follow the market need and usually starts with the standardization of test methods. However, for the purpose of this review, it is the intention to use a more logical order. This will start with terminology or definitions followed by sampling and methods of test. Once test methods are agreed, it is then possible to proceed to classification product standards and specifications for use. Standardization and its progress or lack of it in these areas is reviewed below.

3.1 Terminology

Before any standardization is possible, it is necessary to define the terms to be used. For these 'new' ceramics, the main obstacle is to decide on a suitable adjectival description. Possible contenders (in alphabetical order) are; advanced, engineering, fine, high performance, industrial, special, structural and technical. The USA has opted for 'advanced' and Japan for 'fine'. At present there is no consensus in Europe and this needs to be resolved at the forthcoming meeting of CEN/TC 184*. With regard to the bulk of the terminology, good bases will be available from the completed work in ASTM C-28 and the work in progress in the BSI.

3.2 Sampling

At present, no national standards body is currently working on this subject specifically for advanced ceramics, although it is on the provisional work programme of CEN/TC 184. Standard sampling plans for inspection by attributes are already available as ISO 2859, IEC 410

*The term agreed at the CEN/TC 184 meeting is 'Advanced Technical Ceramics'. This will be accepted throughout West European countries.

TABLE 1
On-going activities in standards for engineering ceramics

General area	Specific area	D	F	UK	US	J
Nomenclature				+	+	
General textural properties	Cracking		□	○	+	
	Density/porosity	○		○	+	
	Grain size	△		+	△	
	Surface texture	△	□	+		
Specimen preparation	Specimen shape		□			
	Surface preparation	△	□			
Mechanical properties room temperature	Short term strength	+	□	+	+	○
	Elastic properties	+	□	+	△	○
	Fracture toughness	+	□	△	△	+
	Fatigue	+				
	Static fatigue			△	△	
	Hardness	+		+	△	+
	Residual stress	△				
Mechanical properties high temperature	Short term strength: bend/tensile	△	□	△	△	○
	Elastic properties	△	□	△	△	○
	Impact					
	Fracture toughness	△	□			+
	Creep	△		△	△	+
	Fatigue	△				
	Low load deformation			○		
	Crack growth					
	Thermal shock resistance	○				△

Physical properties high temperature	Thermal expansion	○		○		△
	Thermal diffusivity		□	○		+
	Thermal conductivity	△				+
	Specific heat			+		+
	Permanent change in dimensions			△		
Electrical properties	Dielectric constant	○	□			
	Electrical resistance	○				
Corrosion	Oxidation			○		
	Aqueous electrolytes	○		○		
Erosion						
Wear						
Adhesive joints		○				
Powders	Particle size		□		△	
	Specific surface		□		△	
	Compressibility	○	□		△	
	Rheology		□		△	
Processing	Green body density	○	□		△	
	Sinterability		□		△	
Analysis	Structural		□		△	
	NDE					
	Chemical		□			○
	Statistical		□			○

○, completed standard; +, activity in progress; △, future activity; □, under consideration.

and BS 6001. Plans for inspection by variables are given in ISO/DIS 3951 and BS 6002. The Electronic Components Committee of CENELEC (CECC), has also produced sampling plans for inspection of electronic components by attributes as part of its 9000 series on assessed quality. It is anticipated that CEN/TC 184 may decide to develop specific sampling plans for such advanced ceramics as composites, coatings, powders and fibres.

3.3 Test Methods

The standardization of test methods is the most advanced of the standardization subjects. However, despite this, there is much work to do especially in the field of mechanical and thermo-mechanical behaviour. The situation is reviewed below under seven headings.

3.3.1 Chemical and mineralogical analysis

Conventional chemical analysis methods are standardized for oxide ceramics. Standardization also exists for methods involving physical methods of analysis such as X-ray fluorescence. These are mainly intended for oxide ceramics. Generally speaking, the chemical analysis of non-oxides in isolation presents no problem. Difficulties can arise, however, in the speciation of mixtures which can contain one or more of the following: oxides, carbides, nitrides, metallic elements, free carbon and oxynitrides. This is a problem which is international and could possibly be resolved by international co-operation.

3.3.2 General and textural properties

Subjects under this heading include methods of test for the determination of the presence of cracks, density and porosity, grain size and surface texture. Of these, surface texture is an important property as it is used to characterize test pieces for mechanical testing. It is normally measured by the stylus method and the equipment is fully standardized. However, this equipment is principally designed to measure the surface roughness of metals and it may be necessary to set out different procedures for ceramics.

3.3.3 Mechanical properties at room temperature

This is the area which is receiving the greatest attention. Property measurements which require to be standardized include: short-term strength, modulus of elasticity (and Poisson's ratio), tensile strength, fracture toughness, fatigue, delayed failure, hardness and wear resistance.

Much work has been done on flexural strength and standards are available. It is important in the international sense that these methods are directly comparable. There are many test methods which are standardized for the determination of fracture toughness. These are mainly intended for metals and are difficult to adapt for ceramics. It is anticipated that hardness and wear resistance will be resolved through the VAMAS programme.

3.3.4 Thermo-mechanical properties

These include some of those properties which are to be standardized at room temperature. In addition it also includes the determination of creep (and sagging) and the resistance to thermal shock. Creep resistance is an important engineering parameter, particularly in the selection of materials for turbine blades. The philosophy of testing thermal shock resistance should be similar to that for wear testing, i.e. to try and simulate practice.

3.3.5 Thermo-physical properties

This includes the determination of thermal expansion, thermal conductivity, thermal diffusivity and specific heat. The favoured way of measuring diffusivity is by the laser flash method.

3.3.6 Corrosion and oxidation resistance

Because of the wide range of chemical environments which may be used to contain ceramics, it would seem sensible to standardize a general guidance on appropriate corrosion test methods. It is, however, appropriate to have a specific standard test for the determination of oxidation resistance.

3.3.7 Characterization of fine powders

The main characteristics are particle size and its distribution. These are largely standardized both in the ASTM and ISO. However, difficulties may be found with ultra-fine powders and the particle morphology. This may require resolution by further study and work.

3.4 Classification

Once test methods are established, it is then possible to classify the various ceramic products. Classification is of commercial importance as it enables comparisons to be made between products from different suppliers. The need for classification has been recognized by the

VAMAS Steering Committee, who have approved the initiation of a pre-standards activity to support the development of a unified classification system for advanced ceramics. The objective of this work programme is to identify and assess the issues inherent in developing a unified classification system for advanced ceramics; establish a building block structure featuring critical elements necessary for international use and establish mechanisms and institutional links as needed and appropriate, between national standards bodies for further system development and refinement to meet individual national and international industrial needs.

A working party, representing participating countries and standards groups will concentrate on three areas: (1) identification of existing classification schemes for parallel, similar or other materials classes amenable to advanced ceramics, assessing attributes and difficulties; (2) establishment of a priority hierarchy of development pathways; and (3) development of a conceptual classification and terminology critical structure, by adaption or through new approaches and determination of the feasibility of a single system for both market indicators and technical elements, considering the following classification factors:

A. Organization by end product and use categories.
B. Categorization by major chemical components.
C. Sub-grouping to infer functional attributes.
D. Coverage of major product forms.
E. Use of consistent terminology.
F. Incorporation or property/performance regimes and validation procedures.
G. Compatibility with computerized data bases.

A 2/3 year effort is envisaged with the work carried out in serial fashion, roughly divided according to the technical approach elements given above.

3.5 Specification and Product Standards

Classification is established to provide the broad framework for the detailing and setting up of specifications for use and product standards. It is anticipated that the standardization of these specifications will be the responsibility of user groups rather than material groups. Standard specifications already exist in the electrotechnical field and the concept is being examined in other areas.

4. CONCLUSION

The commercial exploitation of the new advanced ceramics is poised to take off, and this has stimulated a need for product standards. This need is on an international basis led from three viewpoints, Europe, Japan and USA. There are dangers of duplication of effort and also of national prejudice. It is essential to have openness of discussion and good international co-operation to ensure that positive progress is made.

ACKNOWLEDGEMENT

The author is pleased to thank Dr D. W. F. James, Chief Executive, British Ceramic Research Ltd for his permission to present and publish this paper.

3

The Commercial Development of an Engineering Ceramic

R. J. LUMBY

103 Solihull Road, Shirley, West Midlands, B90 3HW, UK

1. INTRODUCTION

The development of an engineering ceramic can be divided into a number of easily identified stages. The early laboratory development of composition is followed by the optimising of composition, structure and properties. The investigation of the performance of the novel material in a working environment requires as much care and perception as the earlier stage of development.

2. MATERIAL DEVELOPMENT

The development of a material requires a study of composition, structure, properties, and the establishment of trends in property behaviour. Compositional limits can then be defined which allow a material to be prepared consistently to a set combination of property characteristics. In this way an optimum choice of material can be made. This is frequently a compromise between composition and processing, in order that the best combination of properties and cheap processing can be obtained.

3. MATERIAL EVALUATION

The investigation of the performance of a novel material deserves to be more than just a material substitution. In addition, the investigation should not be restricted to one attempt, and must be an evaluation of the material, as opposed to the way the material is applied. The exploitation stage of a novel material has often been estimated to cost *ten* times more than the earlier development stage, which may explain why this stage of development is either neglected or at best done in an unsystematic way. Novel material, by definition, may well exhibit some novel mode of behaviour, which any evaluation must seek to identify. An investigation of the performance must identify both the negative as well as the positive features of the behaviour of the material so that shortcomings as well as advantages are understood. A more systematic approach can result in identifying unexpected behaviour which can range in magnitude from a 100 times increase in life to an unanticipated short life resulting from an unexpected 'Achilles' heel'. It is vital that either of these cases be identified, from a commercial standpoint. In the case of the former, such a large advantage in performance allows a more extensive and 'fail safe' evaluation programme to be afforded. Early identification of an 'Achilles' heel', once and for all lays to rest a possible myth that a particular material can operate successfully in a particular environment. Most importantly, it prevents further waste of time, resources and money.

4. MARKET EXPLOITATION AND ITS DIFFICULTIES

A common practice which must be avoided wherever possible is the mixing of commercial business with the evaluation stage. Whilst with good luck and judgement it can lead to rapid success, it can also lead to problems. One possible result is that the suppliers' state of ignorance is aired in public, which causes unnecessary embarrassment, and probably compromises future business.

It is hardly surprising that the 'suck it and see' approach to the evaluation stage frequently fails. An unsuccessful 'one off' adds little added understanding, and often provokes wild speculation (based on no information) on the cause of failure. In addition it often precludes any future opportunity for further investigation. Careful preparation of evaluation conditions, both of the equipment and of the people involved, is necessary to maximise the yield of trial data. The expense involved is the trial itself, and not the organisation of the evaluation procedure.

The more systematic approach is of course invariably unpopular, for a variety of reasons. At the head of the list is cost, to both the supplier and user. The material developer can rarely afford to spend time and money developing a cost effective process for a reluctant customer. After a successful 'one off' the user can become flushed with the unexpected success but can become very protective of the advantage which he may scarcely deserve. He may expect the supplier not to deal with any of his competitors. As time proceeds the user may well increase his orders, but can often become very cost conscious, and be able to negotiate discounts in exchange for increased numbers. The net result is that the supplier works much harder to be as profitable. At the stage when a successful evaluation ought to allow a more expansive approach, both supplier and user can become rather too defensive. In such an atmosphere, it is hardly surprising that performance is neglected, and cost becomes paramount. The net consequence of such a scenario is that long term priorities are neglected. The evolution of cheap production processes and the development of guaranteed quality are neglected. The customer in the end is the loser.

The avoidance of this sequence of events will always demand a special combination of people and circumstances. It will require an unusual level of understanding and commitment on the part of both supplier and customer. There has to be a measure of understanding of the material, its performance, and the environment in which it is to function. It requires that the customer be committed to the establishment of a spectrum of understanding of performance, and that he be willing to make time available. He needs to be interested beyond considerations of cost and the first possibly misleading result.

5. THE DEVELOPMENT OF SIALON MATERIALS AND METAL CUTTING

Some of the requirements for a commercially successful product are that a customer has a problem product or process and that one material can offer a unique solution to that problem.

Sialon materials had been shown in 1977[1] to have the ability to cut cast iron and tool steel to good effect. Rolls Royce at Derby, at that stage, wanted to retool their turbine disc machining facility with CNC machines with a higher speed capability. At that time tungsten carbide cutting tool inserts were used but would operate at surface cutting speeds of only 15–20 m/min on the disc alloy material Incalloy 901. It was quickly established that alumina ceramic inserts could possibly raise cutting speeds to 100 m/min but with limited tool life, whereas sialon inserts could possibly raise cutting speeds to 250–300 m/min.

One key issue was to identify the characteristics of material composition, structure and properties which would result in predictable and adequate tool life.

5.1 The Characteristics of the Sialon Materials Investigated

The sialon materials at that stage were β'-sialon/glass in structure, were fully dense, and had high strength. The development programme had been controlled by a target strength, measured as a bend strength, of 800 MPa.

Materials were prepared from silicon nitride powder, yttrium oxide (Y_2O_3), and an addition of aluminium and nitrogen in the form of the 21R polytype of aluminium nitride.

The sintering and reaction process, by which the powder preform is converted to a fully dense, strong product, deserves further explanation. The primary function of the Y_2O_3 addition is to induce the formation of the liquid phase by which liquid phase sintering and densification can occur. In the early stages of development it was found that adequate levels of density and strength required a minimum addition of 6% Y_2O_3. The Y_2O_3 reacts with the silicon dioxide (SiO_2) surface layer present on all Si_3N_4 powders.

$$2Y_2O_3 + 3SiO_2 \longrightarrow Y_4\,Si_3\,O_{12}(\text{eutectic liquid})$$

During the sintering process at temperatures between 1550 and 1650°C this eutectic liquid forms allowing particle rearrangement, shrinkage and densification to occur. The mainly α-Si_3N_4 particles in the powder dissolve in the liquid and reprecipitate as β-Si_3N_4. The Al and O present partition themselves between the β-phase and the eutectic liquid. This results in an expansion of the β-phase. It is easy to compute the approximate proportions and composition of the liquid, and therefore that of the resultant glass phase in the sintered product. For a 6% Y_2O_3 addition, and for a typical SiO_2 content of 4%, there will be about 12% glass by weight, which will be approximately 10% by volume (based on the assumption that the density of the glass is 3·0 Mg/m^3). It should be evident that without the measurement of the oxygen content of the nitride constituents and its careful control, the composition and volume fraction of the second phase, the glass, can vary considerably.

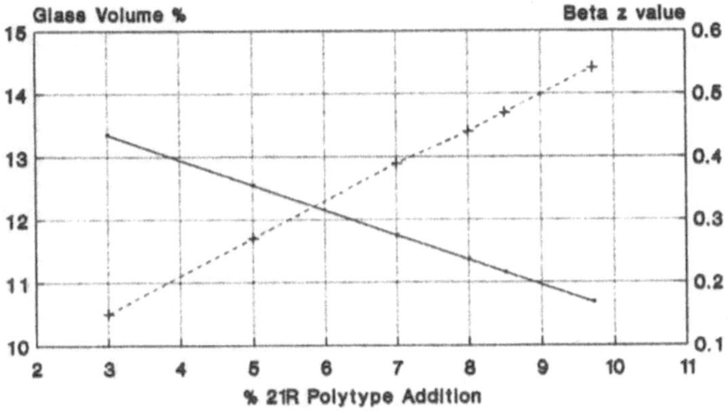

Fig. 1. Effect of polytype additions on glass and beta-phase. (——) Glass volume, (· · · · · ·) beta z-value.

5.2 The Effect of the Addition of extra Aluminium and Nitrogen

21R Aluminium nitride polytype has the approximate composition $6(AlN)\cdot SiO_2$ ($SiAl_6O_2N_6$). Increasing the amount of 21R has two obvious effects, a reduction in the glass phase content, and an increase in the Al and O content of the β-phase (i.e. z-value) (Fig. 1). These effects were first observed by X-ray diffraction and by electron microscopy.[2] The chemistry of these changes stems from the reaction between the added Al and N and the SiO_2 in the liquid phase.

$$SiAl_6O_2N_6 + 2SiO_2 = 1{\cdot}5\ Si_2Al_4O_4N_4(z = 4 \text{ expanded } \beta'\text{-}Si_3N_4)$$

The effects of these compositional changes on the mechanical properties of β'-sialon materials have been described.[3] There is a gradual improvement in the room temperature mechanical properties up to a maximum, but there is a much more significant improvement in the high temperature properties. Additionally it is found[4] later that there is a significant increase in the hardness of the materials over this range of compositions and structures (Fig. 2).

5.3 Variation in Performance with Composition, Structure and Properties

The range of materials illustrated above was evaluated in metal cutting.[5] Cutting inserts to the ISO specification SNGN12-04-16 were prepared with a 0·05 mm wide 20° chamfer. A cutting speed of 310 m/min and a feed rate of 0·18 mm/rev had been previously identified as optimum for sialon materials on Incalloy 901 (Fig. 3).

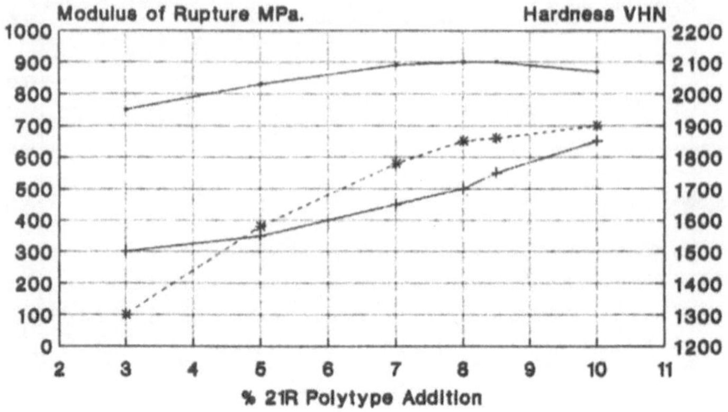

Fig. 2. Effect of polytype additions on strength (25° and 1300°C) and hardness. (—●—) Strength at 25°C, (—+—) strength at 1300°C, (--*--) hardness VHN.

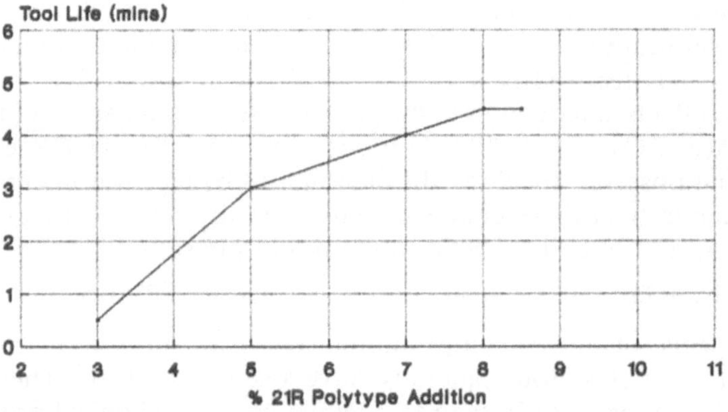

Fig. 3. Tool life of sialon inserts on Incalloy 901.

The above data were obtained from a comparison of the flank wear of inserts of different sialon compositions. Tool life was found to increase with increasing 21R content up to a maximum of 4.5 min for a 21R content of 8%. These tool life data can be reinterpreted in terms of glass content, β'-sialon z-value, hardness, or even the high temperature characteristics.

It would be easy to jump to a conclusion that the improvement in tool life was the result of any of the individual changes which occurred. It is more prudent, however, merely to recognise that these changes are associated and deserve further investigation. Ideally each change should be individually investigated wherever possible.

5.4 Failure Analysis

Further improvements can be obtained by a determination of the cause of failure. High speeds and high temperature often lead to chemical reactions between elements in the work piece and those in the insert material. Identification of the reactions[6] which occur can often lead to remedial treatment.

6. SUMMARY

An attempt has been made to describe the worst aspects of the evaluation process which can occur in the development of an engineering ceramic material. The hindsight provided by an unsuccessful test does not provide recompense for lost opportunity. The evolution of sialon materials as metal cutting inserts, being a successful development, demonstrates perhaps, some of the necessary ingredients of a successful evaluation.

REFERENCES

1. Lumby, R. J., North, B. and Taylor, A. J., Properties of sintered sialons and some applications in metal handling and cutting. In: *Ceramics for High Performance Applications II*, Brook Hill, 1978, pp. 893–906.
2. Butler, E., Lumby, R. J., Szweda, A. and Lewis, M. H., Syalon ceramics for high temperature engines: an illustration of grain boundary engineering. 1st World Conference on Ceramics in Engines, Hakone, Japan, 1983.
3. Lumby, R. J., The preparation, structure and properties of commercial sialon ceramic materials. *Ceram. Engng Sci. Proc., Am. Ceram. Soc.*, **3**(1–2) (1982).
4. Lumby, R. J., The variation of hardness in Lucas sialon ceramics. *J. Mater. Sci. Letters*, **2** (1983) 345–6.
5. Jawaid, A., The cutting of metals with sialon ceramic inserts. PhD Thesis, University of Warwick, UK, 1983.
6. Bhattacharyya, S. K., Jawaid, A. and Lewis, M. H., Behaviour of sialon ceramic tools when machining cast iron. 12th NAMRI Conference, Michigan Technical University, May 1984.

4

Hybrid Silicon Nitride Bearings

S. A. HORTON

SKF Engineering and Research Centre B.V., Postbus 2350, 3430 DT, Nieuwegein, The Netherlands

1. INTRODUCTION

Bearing designers have long been interested in ceramic materials both to increase the performance of bearings and to extend the range of operation to higher temperatures and corrosive environments. Although early work was focussed on bearings made completely from ceramics for high temperature applications such as gas turbine engines, recent development effort has concentrated on hybrid ceramic bearings — steel rings with ceramic balls or rollers. Such hybrid bearings are now available commercially for applications in machine tool spindles and other high speed/precision equipment.

In contrast to many other structural ceramic parts, stresses in bearing components are very high and localised at, or close to, the surface. Consequently, materials for use in bearings have to be fully dense, homogeneous and free from defects near the surface. It is only since the development of hot-pressed silicon nitride in the early 1970s that bearing manufacturers have been able to investigate the potential of ceramics in bearings. Subsequent advances in the understanding of the basic science of nitrogen ceramics and the relationships between structure, properties and defect populations have led to the adoption of silicon nitride as the standard ceramic material for bearings.

2. SILICON NITRIDE

Properties of ceramics and bearing steels are compared in Table 1. Although silicon nitride is neither the hardest nor the toughest of the

TABLE 1
Properties of engineering ceramics and bearing steel

Material	Density ($Mg\ m^{-3}$)	Elastic modulus (GPa)	Hardness (GPa)	Toughness ($MPa\ m^{1/2}$)	Failure mode
Silicon nitride	3·2	310	14–18	5–8	Spalling
Silicon carbide	3·1	420	20–24	2–4	Fracture
Alumina	3·9	390	18–20	3–5	Fracture
Zirconia	5·8	210	11–14	8–12	Spalling
Steel[a]	7·8	200	10	>16	Spalling

[a] Standard bearing steel AISI 52100, or aircraft bearing steel — M50 high speed steel.

engineering ceramics, it is considered to have the best combination of mechanical and physical properties for use in high performance bearing applications.

In addition to the high temperature resistance which stimulated initial interest, silicon nitride has other perhaps less obvious properties or characteristics which give performance advantages in hybrid ball or roller bearings.

Low density — With a density of only 40% of that of steels, silicon nitride rolling elements reduce centrifugal effects in bearings and allow higher operating speeds — important for aircraft engine and machine tool bearings.

High elastic modulus — Silicon nitride has an elastic modulus 50% higher than bearing steels. This increases the stiffness of bearings — in machine tool spindles used for grinding or milling.

Low thermal expansion — Typically 30% that of steel which reduces sensitivity to temperature difference between rings and prevents seizure. In hybrid roller bearings this factor also widens the operating speed range.

Low friction — Friction of lubricated silicon nitride balls against steel is 20% of the steel/steel contact value, giving reduced heat generation and the ability to run under conditions of marginal lubrication.

Stability — Silicon nitride is highly corrosion resistant and does not undergo phase transformations at the operating temperatures and loads of hybrid bearings.

Spalling — If silicon nitride rolling elements fail, then like steels they do so by spalling, the least harmful mode of failure.

In addition, the high degree of dimensional quality and surface finish that is achievable with silicon nitride balls gives advantages of reduced vibration, lower noise and higher precision in hybrid ball bearings.

3. MANUFACTURE OF SILICON NITRIDE ROLLING ELEMENTS

It is convenient to divide the manufacture of silicon nitride balls and rollers into primary processing and finishing. The term 'primary processing' is used here to describe all stages in manufacturing from powder to the formation of densified ball or roller blanks; 'finishing' covers subsequent grinding, lapping and honing operations.

3.1 Primary Processing

In the early stage of development of ceramic bearing components, the starting material was hot-pressed silicon nitride which was available in the form of discs or slabs. Ball, roller and ring blanks were then machined from these slabs using diamond tooling as illustrated in Fig. 1 for ball blanks. Such machining operations were time consuming and expensive and obviously not suited for high volume production, with the added factor of poor material utilisation. Clearly, net-shape manufacturing methods are needed for the efficient production of ball and roller blanks.

Figure 2 shows examples of the processing sequence for ball blanks from the original powder to densified blanks. In common with other ceramic materials the mixing/milling of additives is an important stage which influences final product quality. Some form of agglomeration process, for example spray drying, is needed to make the powder flow. Roller preforms can be made most conveniently by die pressing,

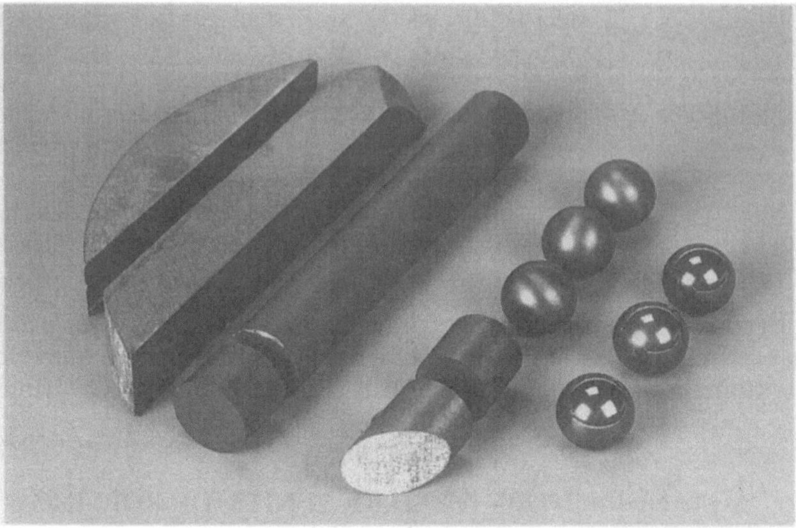

Fig. 1. Balls machined from a hot-pressed silicon nitride disc.

Fig. 2. Balls manufactured via the net-shape route.

whereas there are several methods of producing ball preforms — cold isostatic pressing, injection moulding, die pressing, and compression moulding. Ball preforms can be soft machined in the green state to improve dimensional quality.

Densification by pressureless sintering tends to produce dense material, but usually with some residual porosity which leads to poor rolling contact fatigue resistance. Hot isostatic pressing (HIP) is now seen as the preferred method of manufacture for bearing components since this process, applied directly to encapsulated preforms or as part of a sinter/HIP method, produces 100% dense material.

3.2 Finishing

Balls can be finished in the same way as steel balls in conventional ball lapping machines. However, due to the high hardness of silicon nitride it is necessary to use diamond abrasives for these operations. With steel balls, the excess stock is removed by soft grinding (before heat treatment) and hard grinding (after hardening) using large silicon carbide abrasive discs. These grinding steps are not feasible for silicon nitride balls, and lapping times are longer than for steel balls. Due to the isotropic nature of modern HIP or sinter/HIP materials in comparison with hot-pressed material, it is possible to produce finished silicon nitride balls with better dimensional properties and surface finishes than steel balls.

Roller finishing is more complicated since at least eight different operations are required and these involve machining with fixed abrasive tooling. Furthermore, although rollers have a superficially simple shape, both the ends and the outer diameter can be crowned with 'drops' measured in micrometres. Roller machining operations are line processes unlike the batch finishing of balls, so that care has to be taken to compensate for tool wear to ensure that the correct sorting tolerances are met.

4. MATERIAL QUALITY

Unlike steels, there are no material or composition specifications for ceramics such as silicon nitride; nor are there standard methods for determining some of the important properties. It is also true to say that once densified, the 'structure' of silicon nitride is set, whereas bearing steels after refining and casting are subjected to substantial hot working and thermal processing which refines the structure and can lead to healing of defects. Parameters used in the evaluation of silicon nitride material quality for bearing applications include density variation, microstructural assessment and surface quality assessment of finished components.

The low toughness, or alternatively defect tolerance, of modern high performance ceramics is a point of concern to the engineer. However, bearing manufacturers have been working with relatively brittle tool steels for more than 80 years. Potential defect populations in silicon nitride are generally of the same types as would be found in bearing steels (for example inclusions) or in powder metallurgy (PM) and hard metal products. In many instances they arise from similar effects. For example cracks can be formed as a result of high ejection forces or from stiction effects during cold die pressing, as can occur in PM products.

An example of the microstructure of bearing quality silicon nitride is shown in Fig. 3(a), and undesirable constituents — porosity, inclusions and a healed pressing defect are illustrated in Figs 3(b)–(e) respectively. The high degree of surface finish that can be attained with silicon nitride balls may be judged by a comparison of Fig. 3(f) (ball surface) and Fig. 3(a) (polished section of the same type of material). Surface quality of silicon nitride bearing components is of particular importance as this is where stresses are developed during running. Hertzian stresses below the surface can be higher than 2 GPa. Cracks, pits, inclusions and pressing defects at the surface can therefore lead to a dramatic reduction in bearing life.

Non-destructive evaluation (NDE) techniques have therefore had to be developed for the detection, measurement and classification of defects and non-conformities in silicon nitride balls and rollers. Some of NDE techniques which may be used for material and surface quality assurance include:

X-ray radiography. Standard X-ray radiographic techniques are of little use for bearing components due to their low resolution. Microfocus radiography can be effective in detecting metallic inclusions where there is good contrast due to the high density and absorption of metallic phases. However, ceramic inclusion or agglomerations of glassy phase will not be detected.

Ultrasonic techniques. For bearing components, surface wave techniques are capable of detecting fine surface cracks. These involve the use of special probes operating at frequencies of 50 MHz or higher.

(a) (b) (c)

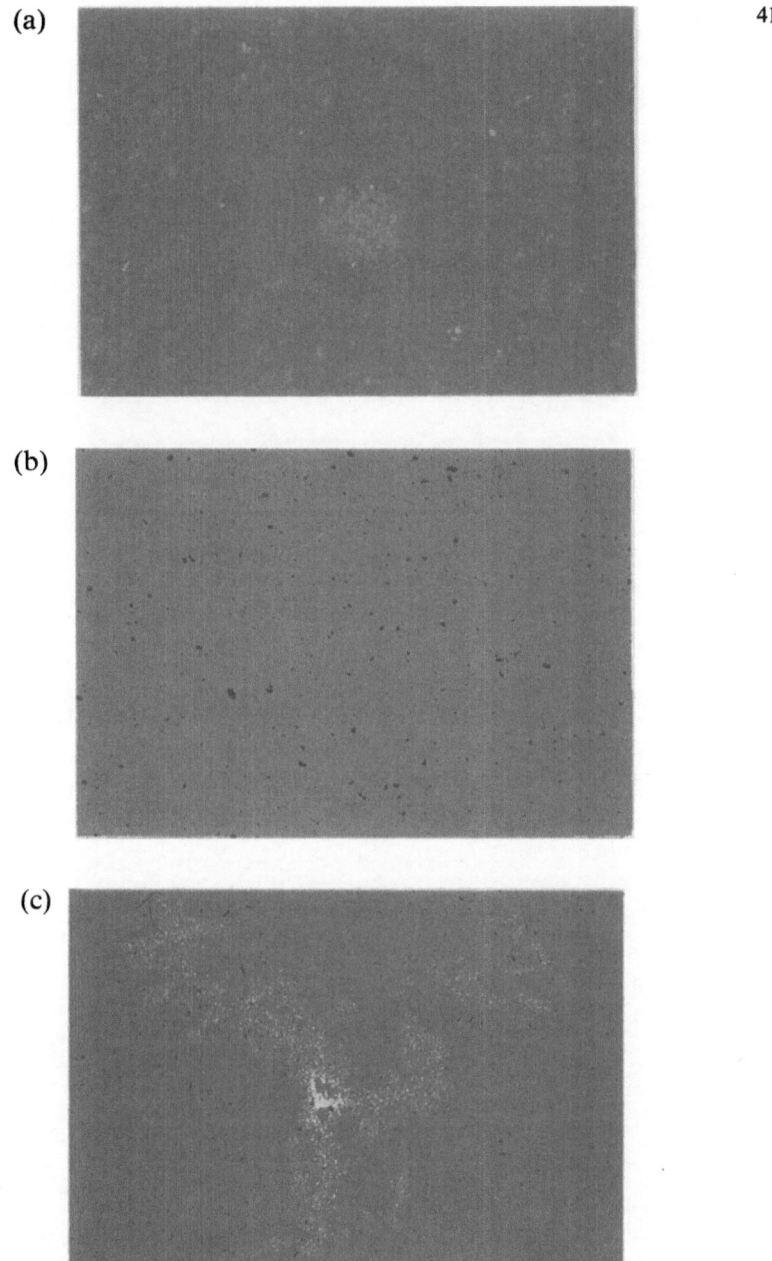

Fig. 3. Microstructure of silicon nitride. (a) Typical bearing grade material (magnification ×500), (b) high porosity (magnification ×100), (c) metallic inclusion (magnification ×100).

(d)

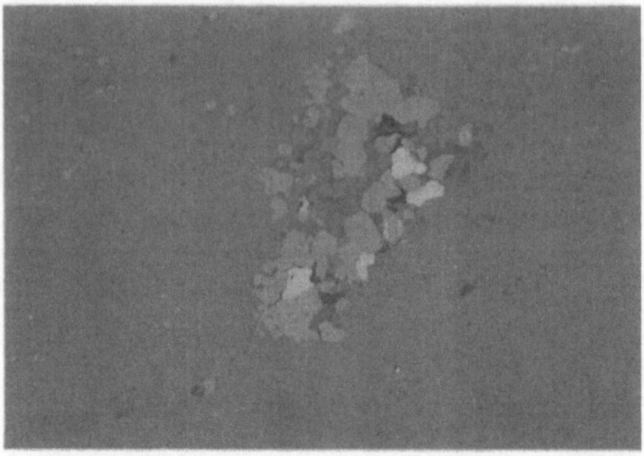

(e)

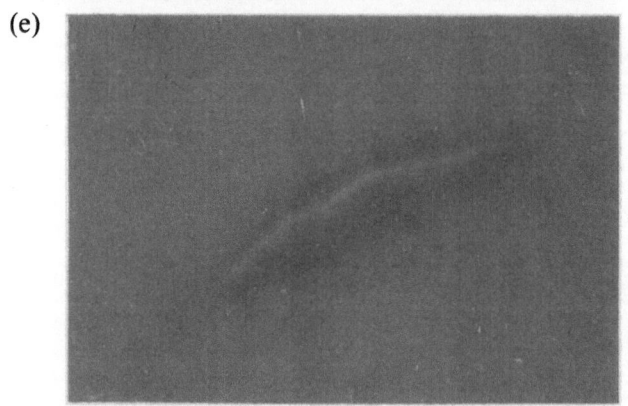

(f)

Fig. 3. *contd.* (d) Ceramic inclusion,(magnification ×500) (e) cold pressing defect healed with glass on ball surface (magnification ×50), (f) finished ball surface (magnification ×500). (a)–(d) are polished sections.

Fluorescent dye penetrants. Although simple in operation, fluorescent dye penetrants are very effective in detecting surface defects on ceramic balls and rollers. Careful control of the processing and the use of specialised examination techniques allow the detection of very fine, optically invisible, cracks. However, some open defects will not retain the penetrant, nor will healed pressing defects (Fig. 3(e)).

Light microscopy. Examination of bearing components using standard brightfield and darkfield illumination conditions is used to supplement fluorescent dye inspection since large inclusions and healed pressing defects are readily distinguished at relatively low magnifications. In addition, examination at high magnifications gives additional information on material quality due to the high degree of surface finish.

In addition to faults present in the material, further defects can be introduced when fixed abrasive hard diamond machining is used. These operations are used in the early stages of roller finishing and the defects formed can be particularly fine. Some examples of machining damage on rollers detected by fluorescent dye penetrant examination are shown in Fig. 4.

5. ROLLING CONTACT FATIGUE

Fully dense and homogeneous silicon nitride has good rolling contact fatigue resistance. As shown in Fig. 5, it outperforms M50 tool steel used for aircraft bearings in accelerated laboratory tests. Although the same applied load was used for both types of material, contact stresses were higher for silicon nitride due to its higher elastic modulus. The figure shows that the 10% failure probability for silicon nitride (L10) is greater than the 50% failure probability for M50. Low levels of porosity and large inclusions can significantly reduce the rolling contact fatigue life of silicon nitride. Results for a sintered silicon nitride containing 0·6 vol.% of porosity are included in Fig. 5.

Full bearing tests carried out by SKF have demonstrated that silicon nitride balls give good performance even under severe conditions of thin film lubrication and high loads. At the same time, such tests have also shown the harmful nature of fine surface Hertzian cracks on the balls.

(a)
(b)
(c)

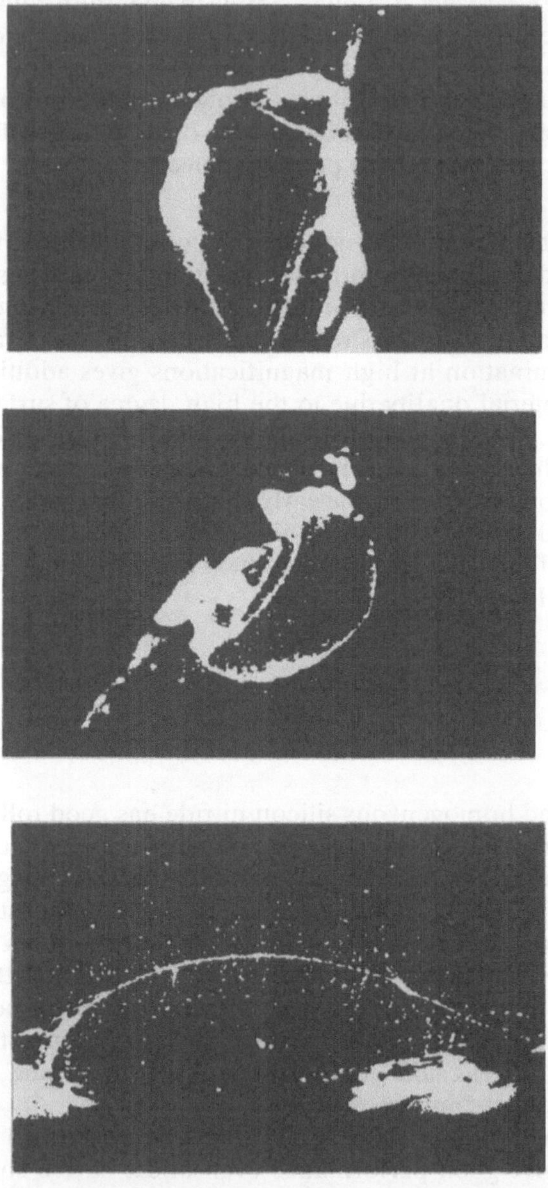

Fig. 4. Examples of defects on silicon nitride rollers, detected with fluorescent penetrant and ultraviolet illumination. Magnification for (a)–(f) × 100. (a) and (b) Incipient corner chips, (c) crack on roller end.

(d)

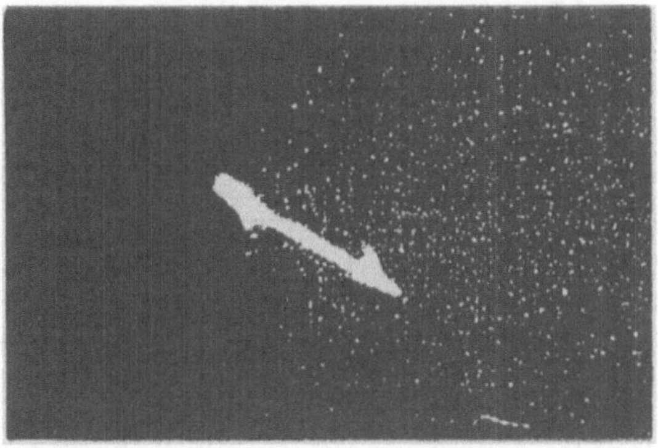

(e)

(f)

Fig. 4. *contd.* (d) Crack extending from edge onto roller diameter, (e) linear array of small micro-cracks, (f) fine multiple crack.

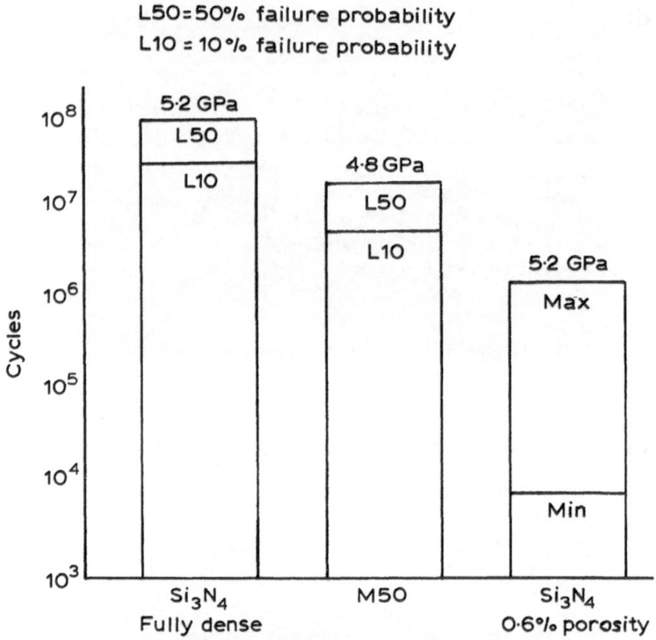

Fig. 5. Rolling contact fatigue resistance of fully dense silicon nitride, M50 high speed steel and porous silicon nitride.

6. HYBRID CERAMIC BEARINGS

The demands on modern bearings have taken the form of steadily increasing operating speeds, greater stiffness and the ability to cope with higher operating temperatures or hostile environments, or a combination of these factors. Although the development of all-ceramic bearings for operation at temperatures up to 800°C remains a long term goal, hybrid ceramic bearings have a more immediate application potential particularly for aircraft engines and machine tool spindles.

6.1 Aircraft Engine Bearings

In aircraft gas turbine engines, the bearings are often the critical factor in limiting the shaft speed that can be achieved. This is due to the increased centrifugal loading on the outer ring as speed is increased and the balls are thrown outwards. (Temperature is not a factor for these bearings since they are cooled by the oil lubrication system). The

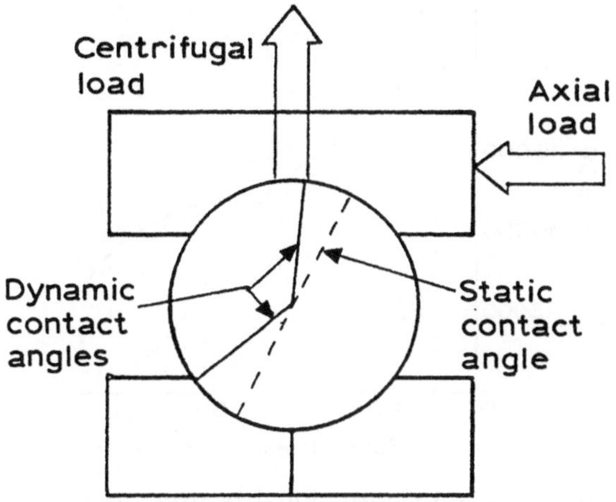

Fig. 6. Centrifugal loading.

situation is shown diagrammatically in Fig. 6, where in addition to increased contact stresses, the centrifugal loading also causes a variation in the contact angles between the inner and outer rings, which in turn leads to skidding and increased heat generation.

Using a helicopter gas turbine engine bearing as an example, Fig. 7 shows the relationship between speed and contact stress. Over the speed

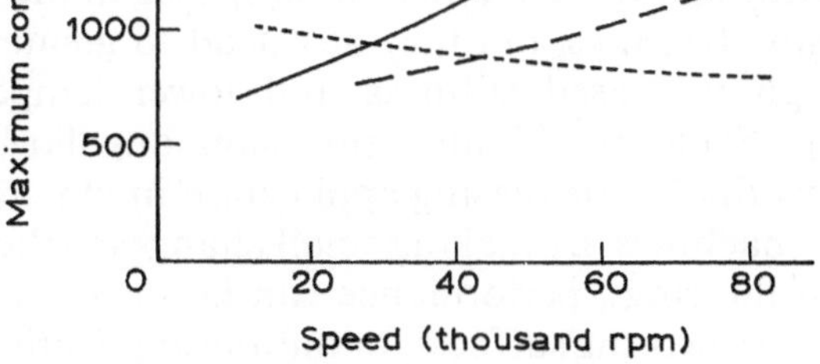

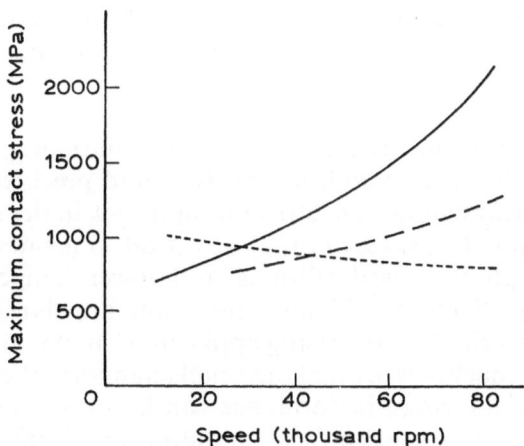

Fig. 7. Contact stresses. (——) Outer ring, steel balls, (---) outer ring, ceramic balls, (· · ·) inner ring.

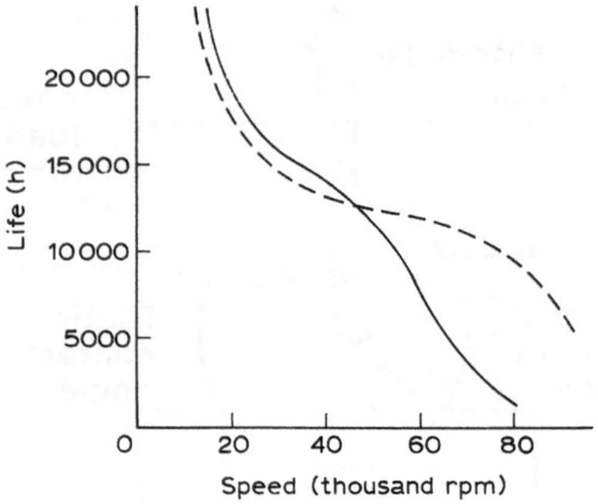

Fig. 8. Calculated life. (——) Conventional bearing, tool steel balls, (----) hybrid bearing, silicon nitride balls.

range of 40 000–80 000 rpm, the contact stress on the outer ring doubles with steel balls, while the use of silicon nitride balls with a mass 40% of steel balls results in much lower contact stresses on the outer ring. The calculated effects on bearing life are shown in Fig. 8. Silicon nitride balls also reduce the contact angle variation (Fig. 6) of the balls with the inner and outer rings which gives better kinematic performance and less heat generation within the bearing.

Reliable NDE techniques to assure freedom from surface defects on ceramic balls is an obvious requirement for aircraft engine bearings.

6.2 Machine Tools

Bearings are also critical components in machine tool spindles as the bearing system influences both productivity and precision. Not only will hybrid bearings allow higher operating speeds in the same way as for aircraft engine bearings, but they also lead to greater machining accuracy through increased stiffness and lower temperature rises during running. Such hybrid high precision angular contact ball bearings are now finding increasing application in machining centres and other CNC machines. By making small changes to the geometry of the raceways in the rings, performance can be further optimised for example to increase bearing stiffness for milling applications (Fig. 9).

Machine tool bearings are usually lubricated by an air–oil mist system since steel bearings will not give the required performance with grease lubrication. Hybrid bearings with silicon nitride balls will give

Series 70 CD
This series is fully interchangeable with the High Speed Range Hybrid bearing and the High Speed Range bearings with steel balls, as the outer dimensions are the same. The properties, however, are greatly enhanced.

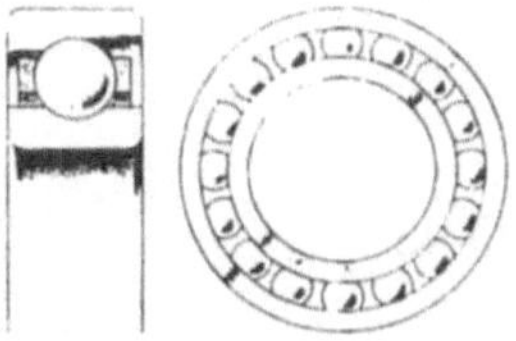

High Speed Range
with steel balls giving high precision and speed.

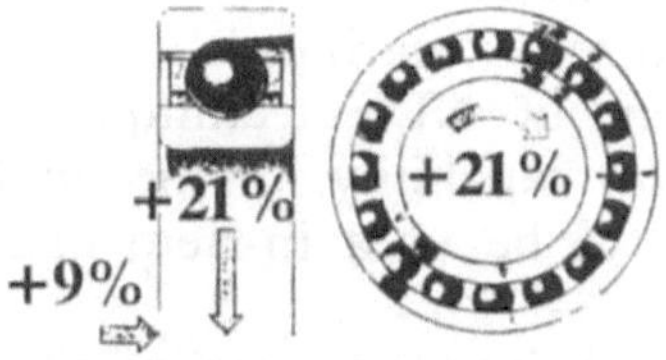

Hybrid High Speed Range
with ceramic balls giving 9% increased axial stiffness, 21% increased radial stiffness and 21% higher speed limits with the same bearing performance as with steel balls

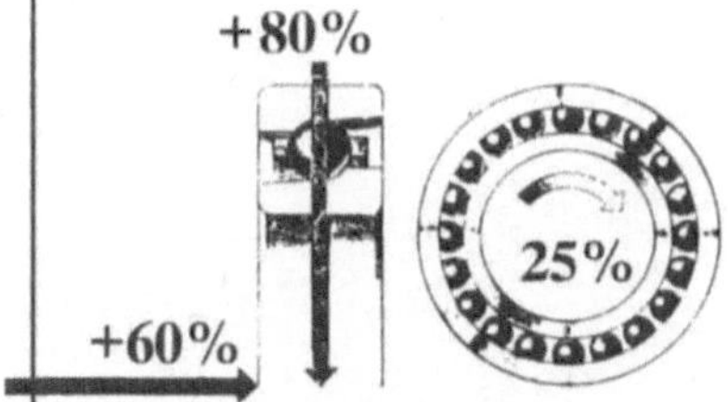

Optimized Hybrid bearing
The Hybrid bearing of series 70 CD with the ultimate bearing performance. It offers extremely high qualities for high precision applications running at high speeds; 80% higher radial stiffness, 60% higher axial stiffness and 25% higher speed limits.

Series 719 CD
This series is fully interchangeable with the High Speed Range Hybrid bearing and the High Speed Range bearing with steel balls, as the outer dimensions are the same. The properties, however, are greatly improved.

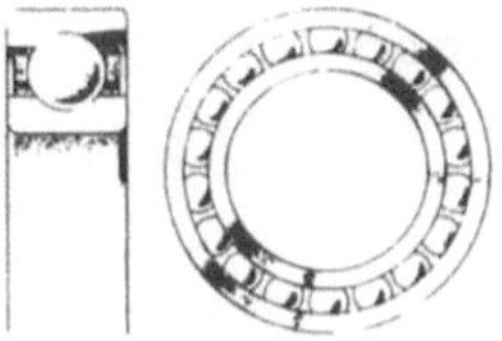

High Speed Range
with steel balls giving high precision and speed.

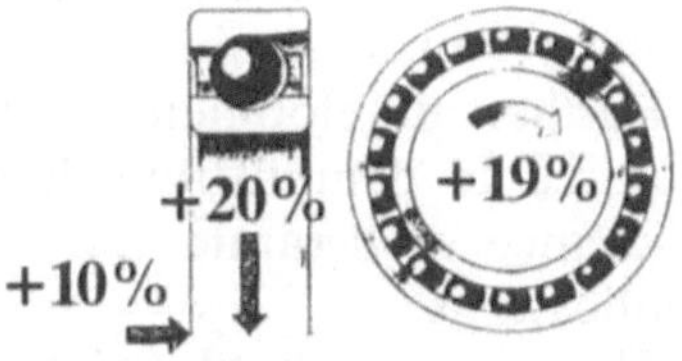

Hybrid High Speed Range
with ceramic balls giving 10% increased axial stiffness, 20% increased radial stiffness and 19% higher speed limits with the same bearing performance as with steel balls

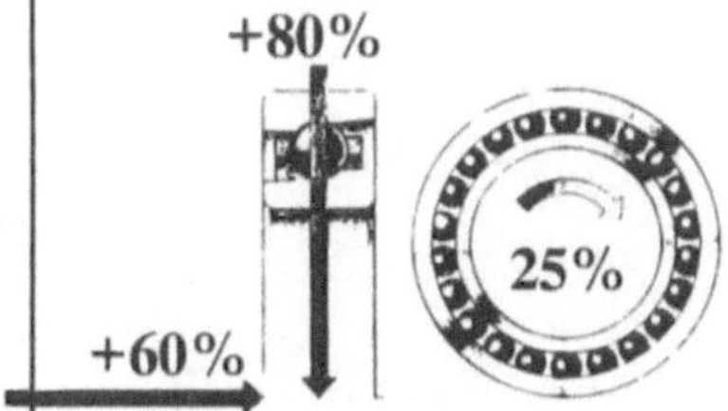

Optimized Hybrid bearing
The Hybrid bearing of series 719 CD with the ultimate bearing performance. It offers extremely high qualities for high precision applications running at high speeds; 80% higher radial stiffness, 60% higher axial stiffness and 25% higher speed limits.

Fig. 9. Comparison of steel, hybrid and optimised hybrid bearings.

the same performance with grease lubrication as all steel bearings lubricated by air–oil mist systems. Changing to sealed grease lubricated hybrid bearings gives a substantial cost saving to the machine builder. Furthermore, in some countries there is legislation against oil–air mists making such a change of bearings inevitable.

Upgrading the performance of machine tool spindles through the use of hybrid ball bearings has provided the impetus for developing hybrid roller bearings, since many spindles contain both ball and roller bearings.

7. CONCLUSIONS

Hybrid ceramic bearings — silicon nitride balls with steel rings — are commercially available, offering significant advantages over their steel counterparts in high performance applications.

Wider application of silicon nitride rollers will increase the overall use of silicon nitride in machine tool spindle units. This is because in some systems a combination of ball and roller bearings is used.

Further development and refinement of NDE techniques are required to demonstrate defect detection capability for the quality assurance of ceramic rolling elements to be used in aero engine applications.

There are probably niche applications for lower quality silicon nitrides and other engineering ceramics where severe environments are encountered and/or contact stresses are low.

5

Product Development with SiC and B_4C Ceramics

PAUL FEINLE and HEINRICH KNOCH

Elektroschmelzwerk Kempten GmbH, Postfach 1526, 8960 Kempten, Federal Republic of Germany

1. INTRODUCTION

Silicon carbide and boron carbide belong to the group of non-metallic hard materials,[1] i.e. materials, whose great hardness and high melting temperature result from a high fraction of covalent bonding. Superhard compounds are formed by appropriate combination of the four low atomic number elements boron, carbon, silicon and nitrogen as indicated by the quarternary system (Fig. 1). Carbon as diamond, boron–nitrogen as cubic boron nitride, boron–carbon as boron carbide and silicon–carbon as silicon carbide, belong to the class of hardest materials known. Further hard members of this system are silicon nitride, the silicon borides and the elements boron and silicon.

In the past, two problems have prevented the widespread use of sintered ceramic components for tribotechnical, wear-reducing applications: (a) a complicated production process marked by relatively high costs and inadequate availability, and (b) the brittleness of ceramic materials,[2] which has kept numerous design engineers from integrating the material into their systems.

In the meantime, progress has been made in both of those areas, that is the technology used for producing sintered silicon carbide (SiC) and boron carbide (B_4C) has become competitive, and much has been learnt about ceramic-oriented designing. There have, however, been but few reports on investigations aimed at clarifying the behaviour of ceramic materials in general, and of sintered non-oxide ceramics in particular, in response to tribological events. Publications mentioned in relevant

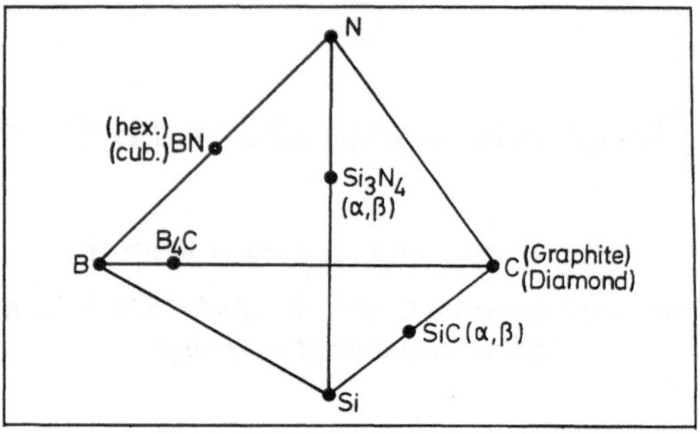

Fig. 1. Combination of elements, B, C, N, Si in 'non-metallic hard materials'.

literature describe the typical material properties of SiC and B_4C and the advantages to be gained from their use. Potential applications for SiC range from the hydrogenation of coal,[3,4] the desulphurisation of flue gases,[5] to the engineering of pumps for corrosive and/or abrasive liquids primarily in the chemical industry.[6-9] They are also used experimentally in high temperature engineering as axial- and radial-turbine rotors.[10]

B_4C components are used, for example, in abrasive wear applications, as sand blasting nozzles or water jet nozzles, and in other special applications such as dressing tools for grinding wheels, light-weight armour plates and in neutron absorber rods and shieldings.[11,12]

Reports concerning theoretical and experimental investigations on SiC appear in the literature, for example concerning the fundamental consideration that the covalent character of the Si-C bond should render it more or less immune to adhesive seizure, SiC thus showing relatively low coefficients of friction.[13,14] Model investigations involving SiC have shown the coefficient of friction to be a function of contact stress,[15] and have documented the anisotropic wear behaviour of monocrystalline SiC crystals.[16] Experimental tribological work has been performed in regard to fretting wear[17] and high temperature behaviour.[18]

In addition to pointing out the wide range of potential applications for sintered SiC and B_4C and listing some practical examples, the present report is intended to show that both the material's properties and its design into real components are important for technical and commercial success.

2. PROPERTIES OF SiC AND B_4C

Because of the relationship between boron and silicon in the Periodic Table of elements, the two carbides SiC and B_4C have some similar physical properties (Table 1) such as a high melting point, very high hardness, high mechanical strength (which remains unchanged even up to high temperatures), a high Young's modulus, low thermal expansion and very good chemical stability. There are relatively large differences, however, with respect to thermal conductivity, resistance to thermal shock, and resistance to oxidation.

Boron carbide has a lower density but similar strength (Table 1), and therefore a higher strength to density ratio than silicon carbide. However, boron carbide is not sufficiently durable at high temperature in air, because it starts to oxidise above 500°C. Its oxidation rate is significantly higher, because the oxide surface layer on boron carbide does not prevent further reaction with oxygen as is the case with the silicon dioxide surface layer on SiC.[19]

TABLE 1
Comparison of SiC and B_4C material properties (hot-pressed, pore-free)

Property	Unit	SiC	B_4C
Density	g/cm^3	3·21	2·51
Melting point (temperature of dissociation)	°C	2 760	2 450
Microhardness	HKO_11	2 600	3 000
Compressive strength	MPa	2 200	2 800
Bending strength (1250°C, Ar, 4 pt)	MPa	400	400
Young's modulus (20°C)	GPa	410	450
Fracture toughness (K_{IC}, SENB)	MPa $m^{1/2}$	3·2	3·0
Electrical resistivity	Ω m	10^3–10^4	10–10^3
Linear thermal expansion coefficient			
(20–500°C)	10^{-6} K^{-1}	4·0	4·6
(500–1 000°C)		5·8	
Thermal conductivity (20°C)	W m^{-1} K^{-1}	110	35
Temperature of application in air (°C), max.	°C	1 600	500
Thermal shock resistance	—	Good	Poor

3. APPLICATIONS OF SINTERED SiC AND B_4C COMPONENTS

Because of the different properties of sintered SiC and B_4C, various application areas have developed. Due to their good mechanical, chemical and thermal properties, SiC parts are used in components for sliding wear, whereas B_4C is successfully used in abrasive wear applications.

3.1 Silicon Carbide

The first large-scale, practice-oriented tests with sintered SiC in the field of sliding wear were conducted on mechanical seals for chemical pumps. Figure 2 shows a selection of rings used in axial sealing and bearing faces of pumps subject to sliding wear.

Consistently positive practical experience gathered on sintered-SiC seal rings[7,8] has helped the material to gain popularity rapidly for use in situations involving wear problems. One of the foremost wear-

Fig. 2. Sintered-SiC rings for axial sealing and bearing faces of pumps subject to sliding wear.

inhibiting properties of sintered SiC is its outstanding chemical stability in any medium, surpassing by far that of hard metals or Si-infiltrated SiC-based materials (SiSiC) containing metallic silicon. Because of its monophase, homogeneous character, it is immune to selective corrosion, even at its grain boundaries. That, together with its high strength, superior hardness, good thermal shock resistance and favourable tribological properties, makes sintered silicon carbide universally applicable for mastering sliding-wear problems. It is also interesting to note that pairing sintered SiC with sintered SiC, for example, in the form of mechanical seal faces, has proved to yield very good results. That advantage is superbly demonstrated by the use of sintered SiC for radially stressed components, e.g. for sliding bearings or shaft sleeves (Fig. 3).

Indeed, sliding bearings using sintered SiC have opened up new technical perspectives: sintered SiC is so physically strong and chemically resistant that such bearings can be positioned directly in the circulating medium, so that the medium itself lubricates the bearing. Pumps equipped with sintered-SiC sliding bearings are being used successfully as standard equipment in the field of chemical process engineering,[8] in areas requiring maximum resistance to corrosion and abrasion.

While most sintered-SiC parts are still used in components subject to

Fig. 3. Sliding bearings and shaft sleeves made of sintered SiC.

Fig. 4. Precision balls made of sintered SiC for use in valves and dosing pumps.

heavy wear and tear, that is mechanical seals and sliding bearings, other areas in which wear resistance is generally necessary are also worth mentioning. These would include corrosion/abrasion-stressed components such as nozzle casings, discharge sleeves and precision parts for valves and dosing pumps (Fig. 4).

3.2 Boron Carbide

Boron carbide is best recognized for its hardness and abrasion resistance.[12] After diamond and cubic boron nitride, B_4C is the third hardest of the technically useful materials. Sand blasting nozzles made of dense sintered B_4C are extremely wear resistant (Fig. 5).

Under highly abrasive conditions B_4C outperforms other hard materials. If hard aluminia grits are used as blasting media for example, then the life-time of a B_4C nozzle is about 100 times better than a hard metal tungsten carbide nozzle under the same blasting conditions (Fig. 6).

Water-jet cutting is a new process steadily growing in importance. Highly pressurized water-jets have been successfully used to cut rock and a wide variety of other materials. In order to improve the cutting performance abrasive particles are added to the water-jet. The abrasives in the water-jet however have a very negative influence on the service life of a water-jet nozzle. Elektroschmelzwerk Kempten GmbH has

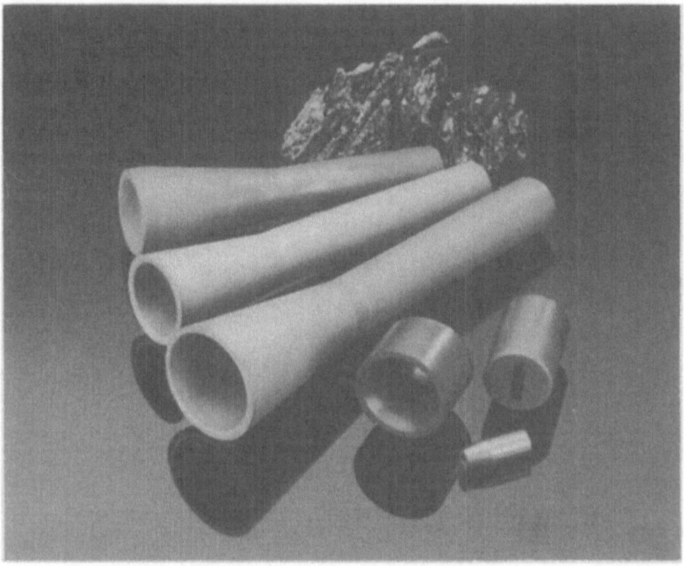

Fig. 5. Sintered boron carbide sand blasting nozzles.

developed a process to manufacture B_4C (Tetrabor®) water-jet nozzles with fine bores, which show good wear resistance against the abrasives loaded water-jet (Fig. 7). Considerable extension of service life improves the profitability of this technology and helps to develop the application potential of water-jet cutting.

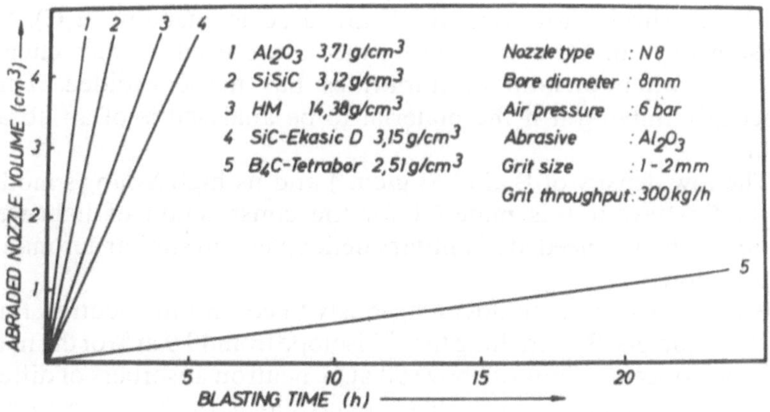

Fig. 6. Comparison of different hard materials in a wear test as a sand blasting nozzle.

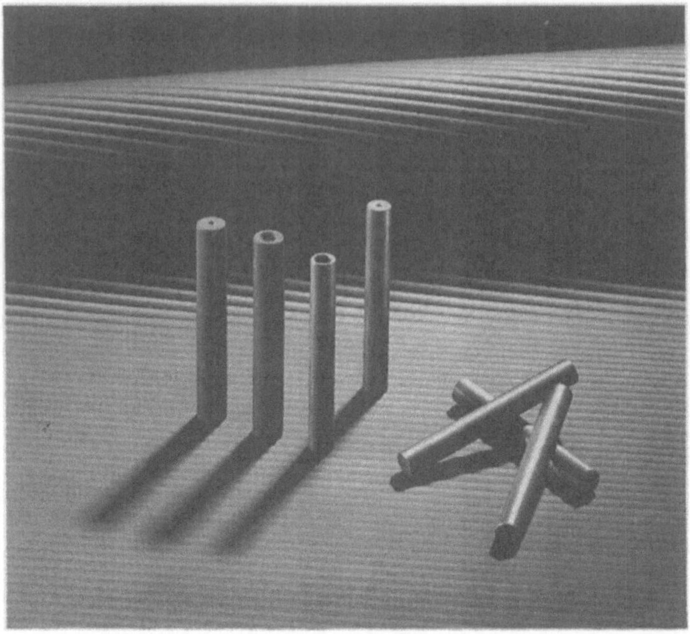

Fig. 7. Sintered boron carbide water-jet nozzles.

The great hardness of B_4C can be exploited also in the form of dressing sticks to contour grinding wheels; this means that sintered B_4C can be used to machine ceramic grinding wheels (it cannot be used for diamond wheels, however, which are used to machine B_4C). Wear resistant B_4C mortars or grinding media are used in trace chemical analysis, when pick-up of impurities has to be avoided. This is especially important if the material to be analysed is of an abrasive nature.

The low density of B_4C (2·51 g/cm^3) and its high Young's modulus (450 GPa) favour this material for the construction of light-weight armour, which is needed in military helicopter and similar aeronautical applications.

B_4C can also be used advantageously to control the neutron flux of nuclear reactors. By enriching the ^{10}B isotope from 19·9 at.% of the natural value to concentrations of above 90 at.%, neutron absorbers of different efficiencies can be produced. B_4C pellets shown in Fig. 8, are normally stacked into neutron control rods. B_4C materials are used to control

Fig. 8. Sintered boron carbide ceramics.

neutron fluxes in boiling water, pressurised water, high temperature, and fast flux breeder reactors. In addition to applications in the reactor core, B_4C based materials (plates, foils, etc.) are used to absorb neutron radiation in, for example, laboratory experiments, and during transport and storage of neutron radiating materials.

4. MATERIAL PROPERTIES AND WEAR

Prior to discussing the materials' microstructural properties and their effects on its sliding-wear behaviour, let us first take a brief look at the

TABLE 2
Chemical composition and corrosion behaviour of 'EKASIC D' sintered SiC

Analysis	Wt %	Corrosion behaviour	
SiC	98·5	Aqueous	No attack
C (free)	1·0	Acids and mixtures	No attack
Al	0·3	Alkaline solutions and mixtures	No attack
Si (free)	0	Organic solvents	No attack
0, N	Traces		

essential causes of wear in everyday practice. We assume that the sliding face is adequately lubricated, thus preventing sustained unlubricated operation. As indicated in Table 2, the chemical stability of sintered SiC protects it against corrosive attack.

Because of its superior hardness and strength, sintered SiC is also very resistant to abrasion by solid particles entrained in the liquid medium. In other words, even with a combination of corrosion and abrasion material maintains good wear resistance.

Many sliding-wear problems occurring in the field are attributable to interruption of the ideal (that is properly lubricated) running conditions, in which case the sliding faces of the bearing or seal in question make contact with each other, giving rise to solid-state or dry friction marked by a pronounced increase in the coefficients of friction. Local frictional heat leads to peak thermal stresses that may be of such intensity as to cause a breakout of microstructural constituents. Then, when lubrication (and cooling) is restored, the material is in danger of cracking or fracturing due to thermal shock. Sintered SiC, however, is better able to cope with such situations than other ceramic materials, because it is stronger and has a lower thermal expansion coefficient, and a higher coefficient of thermal conductivity (cf. Table 1). Consequently, brief periods of dry running can be survived. In fact, sintered SiC even remains stable after prolonged periods of unlubricated operation, although secondary damage may result due to a build-up of frictional heat.

The generalised, typical, sliding-wear phenomena mentioned above, do not of course constitute a complete description. The actual situation within the gap between the sliding faces can only be adequately described by taking into account also the existing forces, pressures, temperatures, velocities and surface structures. In this respect, the reader is referred to specialised literature on complex tribological systems.[20, 21]

Closer attention must be given to the effects of the wear face's surface texture with respect to wear behaviour. In the presence of a hydrodynamic film of lubricant, a smooth, flat sliding face may be assumed to represent the technically most appropriate solution. The situation changes immediately, however, when the lubricant separates or otherwise breaks down, resulting in unlubricated operation, as mentioned above. When that happens, a surface having a specific texture that is capable of providing some form of residual, or forced, lubrication will provide the better sliding properties. Such structures could consist of machined lubricating/cooling grooves, for example, or take the form of lubricant-filled pores, for example, babbit metal, either of which would give off their store of lubricant in the case of dry running, thus preventing, or at least reducing, the occurrence of dry friction.

The porosity of sintered SiC is typically less than 3·5 vol.%, with the pore size situated in the micrometre range. Considering those values, the good emergency running characteristics of sintered SiC can hardly be attributed solely to miniscule pockets of lubricant collecting in the pores. Closer examination of worn surfaces in components after service, shows a more logical explanation for the good behaviour of sintered SiC. Figure 9 shows a polished microsection of sintered SiC, and an SEM micrograph of the worn face of a sliding ring from a sliding ring seal that was taken out of successful service. The worn face shows a sintered-SiC microstructure, in the form of a polished relief surface. Figure 10 shows an enlarged view of the same slide surface; the crystallite-dependent relief structure of the surface is easily recognisable.

It is of particular interest to note that the relief features seen developed in a finely lapped surface under in-service conditions. The relief structure apparently developed as a result of the SiC crystals' anisotropic tribological properties,[16] which are most readily seen with relatively large crystallites. The depressions in such a textured surface can effectively serve as reservoirs for lubricant, thereby improving the component's emergency running properties for situations in which the lubricating film separates and produces a dry-running situation. Figure 11 shows a schematic representation of the lubrication process.

5. DESIGN ASPECTS

The material properties of pure sintered SiC are unique with regard to hardness, temperature stability, abrasion and corrosion resistance.

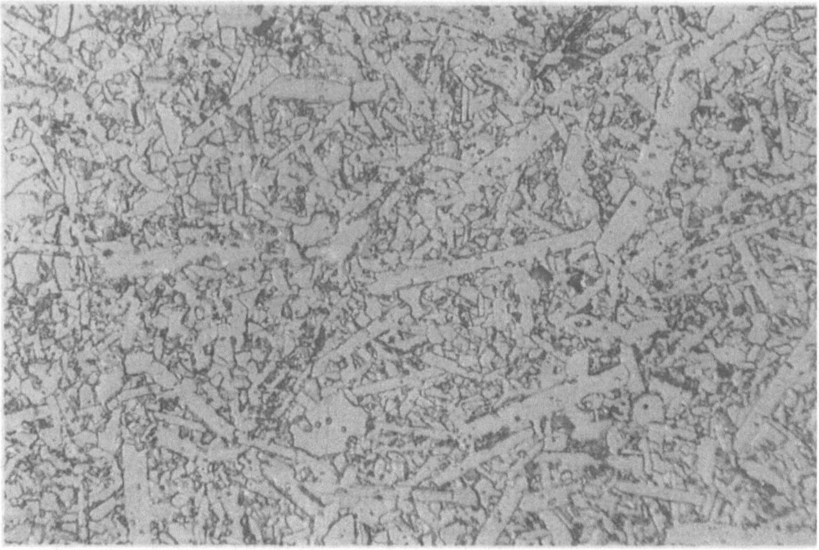

Fig. 9a. Polished microsection of sintered SiC 'EKASIC D'. Scale: 8 mm = 20 μm.

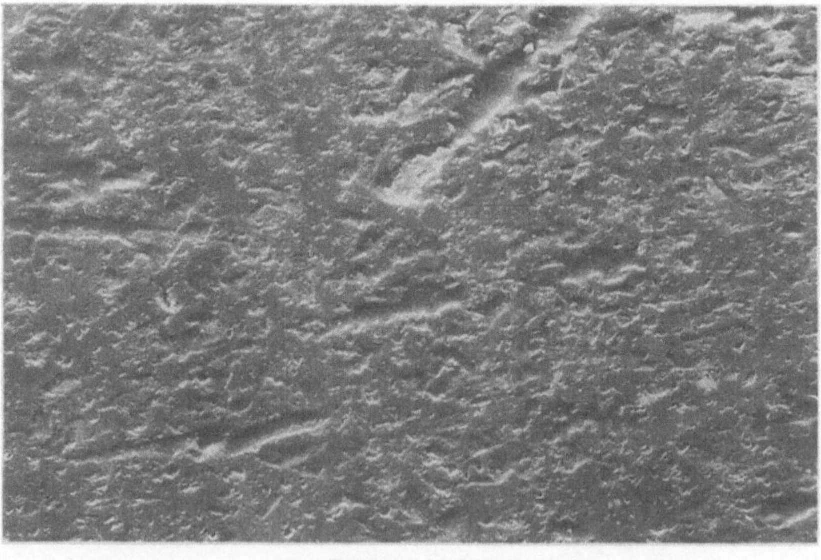

Fig. 9b. SEM micrograph (20 μm) of a worn face made of the same material after approximately 4000 h of service in a mechanical seal lubricated with water; maximum pressure 50 bar, with carbon as friction partner. Scale: 8 mm = μm.

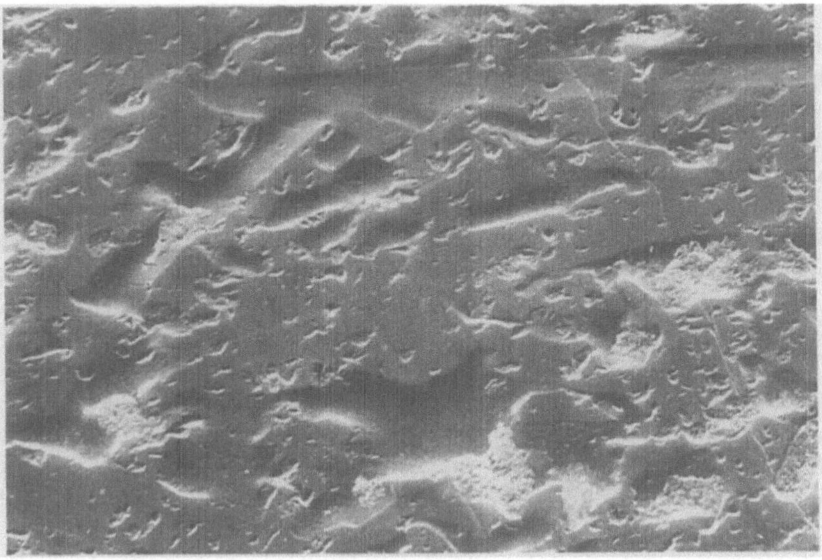

Fig. 10. SEM micrograph of the worn face of a mechanical seal made of sintered SiC 'EKASIC D', otherwise as in Fig. 9b, formation of a surface relief structure. Scale: 8 mm = 6·7 μm.

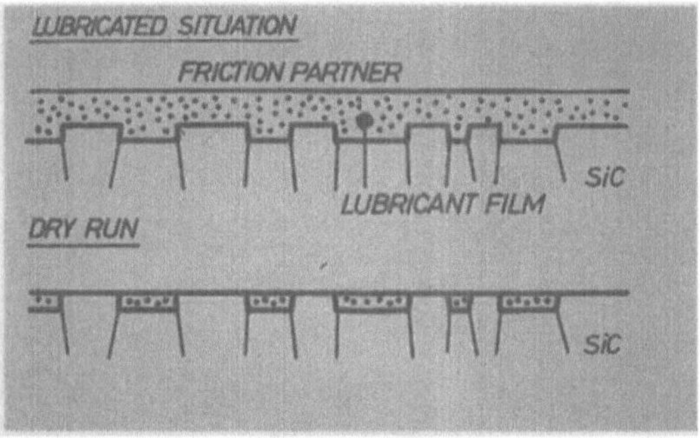

Fig. 11. Schematic representation of the operational situation of a relief-structured worn surface in the lubricated state, and after separation on the lubricating film.

They are the foundation for the universal application potential of sintered SiC wear components used in chemical process engineering; especially when different wear mechanisms coincide; for example, corrosion and abrasion. Besides the many positive properties of sintered SiC, there are some other properties which must be kept in mind to avoid technical problems when hard and brittle ceramic components are used in metallic structures. During the sintering process the as-formed green components shrink in size to the point that the volume of a sintered component is only 60% of that of the green component. This shrinkage means that a very close tolerance can be difficult to control through sintering. Typical as-sintered tolerances are 1% of nominal dimensions; closer tolerances have to be machined. As a consequence of the high hardness of SiC, machining can only be done by diamond grinding. The hardness and low fracture toughness of SiC cause a sensitivity to impact and crack formation. For the construction of components the design has to be correctly evaluated, avoiding especially intensification of tensile stress caused by notches and other stress concentrators. Grinding of sintered components should be kept to a minimum to achieve minimum component cost. The goal is that the design should guarantee the function, at lowest grinding expense. This necessity challenges the designer. A simple example is that in a joint between metal and SiC parts, exposed to temperature changes, the large difference between the thermal expansion coefficients of SiC ($4 \times 10^{-6}\,K^{-1}$) and steel ($12\text{–}15 \times 10^{-6}\,K^{-1}$) has to be taken into consideration. A correctly designed joint of a SiC sleeve to a metal shaft

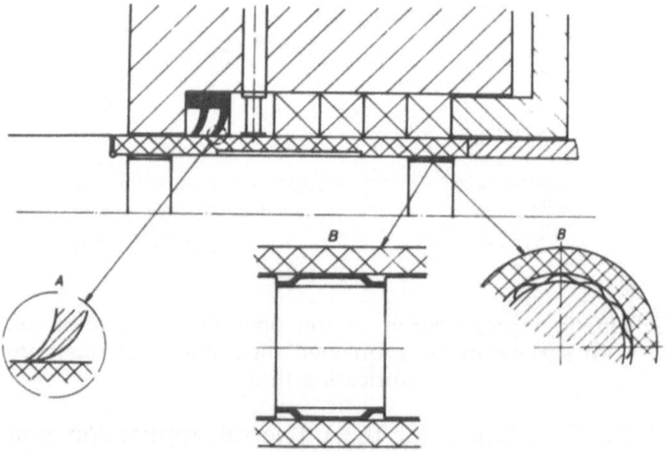

Fig. 12. Schematic of a correct ceramic design: SiC sleeve (cross-hatched) on metallic shaft; (A) area of wear, for example coal filled PTFE rings on SiC; (B) elastic joint between sleeve and shaft by corrugated metal ring.

is shown in Fig. 12, also indicating the critical wear area. The sleeve is fixed to the shaft elastically using corrugated metal rings. The tolerances are such that the metal shaft can expand with temperature increase without coming into direct contact with the SiC sleeve. The correct constructive integration of ceramic engineering components into metallic structures is an essential requirement for successful application of engineering ceramics in the chemical industry. Differences in elastic constants, thermal conductivity and expansion, and the brittleness of SiC (there is no plastic deformation at all) must always be taken into consideration.

A common procedure for joining a ceramic ring to a metal casing is to use shrink fitting. This seems to be a very simple way of joining. However, if we have a closer look at thermal stresses, which are a consequence of different elastic constants, thermal conductivities and expansions, we have to realise that even a simple shrink fit has its kinks.[22] Calculations show that critical tensile stresses, which cause the formation of cracks, may develop in a rather simple shrink fit joint. One possibility for preventing the formation of tensile stresses on cooling is achieved by heating the SiC sleeve and the steel casing to the same temperature before joining and cooling.

For a good design of a sliding bearing, functional and commercial aspects have to be considered. The geometry of the bearing parts should be such, that near net-shape die-pressing is possible, as a low cost forming process. Radial bearing sleeves and bushes must not be too long. A length to diameter ratio of <1 is recommended. Otherwise the different elastic constants of SiC and steel lead to problems if a rotating shaft, carrying the bearing, is elastically vibrating or bending. In order to allow elastic movement of the bearing, an absolutely stiff joint between shaft and sleeve is prohibited. A sliding bearing in a pump normally consists of a radial or journal bearing, taking up the radial load, and a thrust bearing, taking up the axial thrust. For good bearing design it is vital to have journal and thrust bearing mechanically decoupled to allow for some relative movement without creating critical tensile stresses in the ceramic part.

6. CONCLUSIONS

The great usefulness of sintered SiC and B_4C is attributable chiefly to their exceptional chemical, physical and mechanical properties.

Because of their good tribological properties, SiC parts are used in components for sliding wear, whereas B_4C is successfully used in abrasive wear applications. A further factor influencing wear due to sliding stresses is the material's microstructure, which determines the in-service structure of the wear face. Practical experience shows that a matrix structure allowing the formation of a relief texture in the wear face develops the best resistance to sliding wear.

Besides the materials' exceptional properties, good design of engineering ceramic components is necessary in order to make the application of sintered ceramic products a complete success.

REFERENCES

1. Kieffer, R. and Benesovsky, F., *Hartstoffe*, Springer-Verlag, Wien, 1963.
2. Dawihl, W. and Altmeyer, G., Grundlagen des Verschleißes hochharter Werkstoffe, *Wear,* **32** (1975) 291–308.
3. Kelley, *et al.*, Wear resistant materials for coal conversion components, *Proc. Am. Conf. Mat. Coal Conv. Util.*, 1979.
4. Shetty, D. K., Coal slurry erosion of reaction bonded SiC, *Wear,* **79** (1982) 275–9.
5. Schöpplein, W., Flue gas desulphurisation plants-sealing effects, *World Pumps,* **5** (1986) 124–8.
6. Eisner, J. H., Gleitringdichtungen, *Chemie-Anlagen + Verfahren,* **46** (1982) 51, 54.
7. Knoch, H., Kracker, J. and Schelken, A., Bauteile aus gesintertem SiC im chemischen Apparatebau, *Chemie-Anlagen + Verfahren,* (1983) 28–30.
8. Knoch, H., Kracker, J. and Schelken, A., SiC-Werkstoffe für erosiv und korrosiv beanspruchte Pumpenbauteile, *Chemie-Anlagen + Verfahren,* (1985) 101–4.
9. Knoch, H. and Kracker, J., Sintered silicon carbide — a material for corrosion and wear-resistant components in sliding applications, *cfi/Ber. DKG,* **64** (1987) 159–63.
10. Hunold, K., Greim, J. and Lipp, A., Injection moulded ceramic rotors — Comparison of SiC and Si_3N_4, *Powder Metallurgy int.,* **21** (1989) 17–23.
11. Knoch, H., Non-oxide technical ceramics, 2nd Europ. Symp. on Engineering Ceramics, Nov. 1987, London, Riley, F. L. (ed.), Elsevier Applied Science, London, New York, 1989.
12. Schwetz, K. A., Reinmuth, K. and Lipp, A., Herstellung und Anwendung refraktärer Borverbindungen, *Radex-Rundschau,* (1981) 568–85.
13. Glaeser, W. A., Wear and friction of nonmetallic materials evaluation of wear testing, ASTM STP 446, 1969, pp. 42–54.
14. Woydt, M. and Habig, K.-H., Technisch-physikalische Grundlagen zum tribologischen Verhalten keramischer Werkstoffe, *BAM-Forschungsbericht*, 133, Wirtschaftsverlag NW, Bremerhaven, 1987.

15. Richerson, D. W., Contact stress and coefficient of friction effects on ceramic interfaces, *Mat. Sic. Rex.*, **14** (1981) 661–76.
16. Miyoshi, K., Anisotropic tribological properties of SiC, *Proc. Int. Conf. Wear Mat.*, San Francisco, 1981, ASME.
17. Klaffke, D., Fretting wear of ceramics, *Tribology Int.*, **22** (1989) 89–101.
18. Woydt, M. and Habig, K.-K., High temperature tribology of ceramics. *Tribology Int.*, **22** (1989) 75–88.
19. Schlichting, J. and Schwetz, K. A., Oxidationsverhalten von gesintertem α-SiC, *High Temperatures — High Pressures*, **14** (1982) 219–23.
20. Czichos, H., Tribology — *A systems approach to the science and technology of friction, lubrication and wear*, Elsevier, Amsterdam, New York, 1978.
21. Victor, K. H., *Tribologie, Reibung, Verschleiß, Schmierung, Band 10*, Springer Verlag, Berlin, Heidelberg, New York, 1985.
22. Knoch, H. and Sigl, L., Product development with pressureless sintered silicon carbide, In *Proc. of the Silicon Carbide Conference*, Atagawa — Hastsu, Higashizu, Japan, 1989.

6

Oxide Ceramics — Erosion Resistant Materials

C. R. DIMOND

Morgan Roctec Ltd, Bewdley Road, Stourport-on-Severn, Worcestershire DY13 8QR, UK

1. INTRODUCTION

'Wear' defined as the removal of material, is a subject which is known and experienced by everyone. It impacts strongly on the domestic economy as well as industrial and governmental economies. In spite of this the subject receives little scientific attention and the level of knowledge available, when compared with other scientific disciplines, is low.

Very often wear is accepted as a natural consequence of use; 'a fact of life' that cannot be stopped but that can be budgeted for and maintenance schedules written around so as to prevent 'unexpected' breakdowns. Although wear cannot be stopped, the rate of wear can be decreased substantially with the use of the 'best' wear resistant materials correctly installed.

Wear is normally sub-divided into sections to include abrasive, erosive, adhesive and fretting wear and it is erosive wear which is being considered in this paper. Erosive wear is caused by the impingement of high velocity particles which can be either solid or liquid. The velocity of impingement can range in practical examples from 20 m s^{-1} for conveying in lean phase pneumatic systems to in excess of 250 m s^{-1} for radome materials suffering from rain erosion.

One of the materials that has been successfully used to combat erosion and reduce the rate of wear of components in the field of material transportation is sintered alumina and the application of this group of materials only will be considered in this paper.

A recent market research report[1] highlighted the lack of understanding of ceramics by professional engineers and four major concerns were raised:

(i) they are only suitable for high-tech applications and are therefore too expensive;
(ii) they are not impact resistant;
(iii) they cannot be fixed securely;
(iv) they can be used only to effect at high (or low) temperatures;

some of which are addressed in this paper.

2. MATERIAL SELECTION

The choice of class of material is very dependent on the angle of attack of the particles as the relationship between erosion loss and angle of attack is different for different classes of material. For example, with brittle materials the erosion loss increases as the angle of attack increases whereas for ductile materials there is maximum erosion at an impact angle of about 30°.

In many applications concerning particle flow, the actual angle of impact is not known and in practical situations where the conveyed inhomogeneous material has a range of particle sizes and densities, then the angle of impact varies. In most pipe conveying systems impact, and therefore wear, takes place on the outside of the pipe generally at about three or four distinct points around the bend angle, each one covering about 30–60° of the circumference of the pipe. When wear does take place and particularly with ceramic materials, steps can be created thereby increasing the angle of impact. An example of step formed in a 380 mm diameter pulverised coal conveying pipe is shown in Fig. 1. Such changes in local surface profile mean that predicting life of linings is fraught with difficulty and is certainly not an exact science. Despite these obvious complexities, laboratory testing of materials, under a range of conditions, is a worthwhile exercise, not only for comparing materials of one class but also for comparing materials from different classes — although the latter results require substantial analysis before conclusions are drawn.

It is also necessary to take full account of in-service plant trials and use this information to complement the laboratory results so that test methods related to particular applications can be formulated.

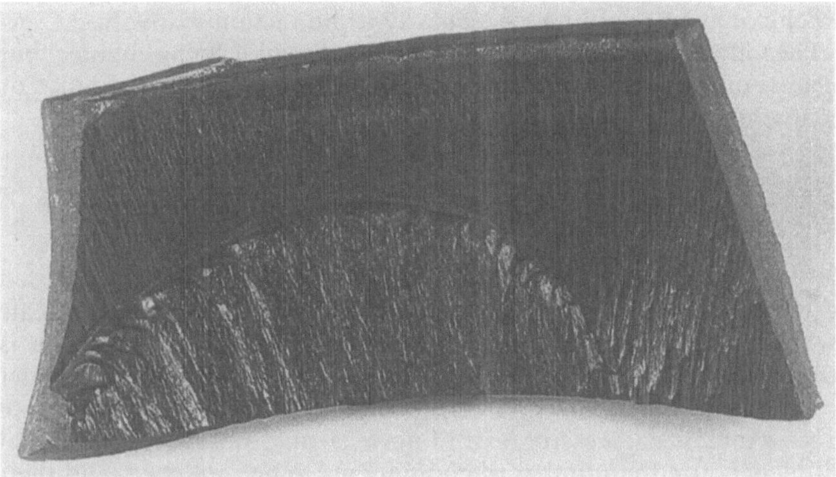

Fig. 1. Wear pattern caused by the flow of pulverised coal.

Other factors that have to be considered in parallel with the materials' wear resistance include:

(i) temperature stability,
(ii) ease of installation,
(iii) cost of materials,
(iv) cost of installation,
(v) surface profile of substrate,
(vi) environmental conditions,
(vii) properties of the erodent.

Various materials such as polyurethanes, rubbers, hardened steels, cast iron, refractories and ceramics have all been used to combat erosion and each has a particular niche as there is not a 'best' material for all applications.

Mechanical and maintenance engineers therefore have to select on the basis of a wear resistance index determined under particular circumstances which may or may not be similar to their own and economic considerations which are based not only on the simple costs (iii) and (iv) above, but on the cost of selective replacement and repair if necessary, cost of spare capacity and the cost due to plant standstill and lack of production. It is certainly a complex decision process, particularly when it is sometimes based on results which are probably not completely relevant and have cost implications, if failure occurs, which could be crippling, particularly if production is involved.

The following information is laid down as a guide to the engineer but because of the nature of the topic it cannot be completely definitive or comprehensive.

3. MATERIAL MANUFACTURE AND PROPERTIES

Oxide ceramics, and in particular sintered alumina, have been used successfully to combat erosion and reduce the wear of large scale industrial plant and the application of this group of materials is considered in this paper. Sintered alumina, which can be classed under the heading 'technical ceramics', can be produced by a variety of process routes, viz. pressing, extruding or casting, and the route selected will depend on the number of components to be produced, and their complexity of shape.

With alumina used in the erosion resistance field, the majority of materials are either die pressed or cast; the casting option being particularly useful and relevant when non-standard complex shapes are being lined, for example, the inside surface of a cone where the cutting of standard flat tiles would substantially increase the total contract value.

The casting option is also of substantial value when relatively large items are required which could not easily, or cost effectively, be manufactured by other process routes. The casting option is also important when small numbers of components are required either for trial purposes or where the component design has not been finalised; tooling in pressing operations can cost around £4000 compared with tooling for casting of around £400.

Figures 2 and 3 show typical 96% alumina cast items manufactured to combat wear in bulk handling operations. Materials manufactured by different production routes have essentially similar mechanical properties so long as full density is obtained and the microstructures are similar. Although this statement might seem obvious, different process routes albeit using the same raw materials can produce different microstructures. For example, additional milling to create a slip of a particular rheological character can lead to breaking down of crystallite size with the result that the alumina becomes more reactive and therefore during kilning can produce excessive grain growth. The importance of microstructure will be discussed later in the paper.

Fig. 2. Cast sintered alumina cyclone nozzle.

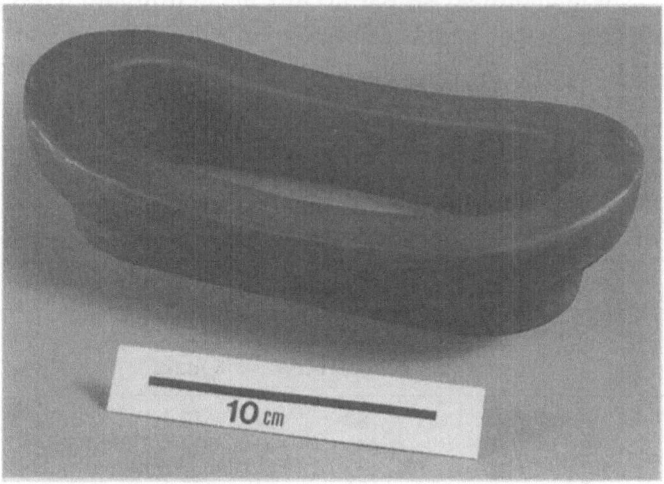

Fig. 3. Cast sintered alumina inlet port.

Despite the apparent equality of alumina ceramics in terms of basic mechanical parameters, it is found in both laboratory wear tests and in real situations that apparently similar aluminas can perform differently by factors of up to eight times in erosion situations.

Erosion testing and particularly erosion testing using 'soft erodents', has been shown to be an excellent selective comparator of aluminas and also as a test to enable changes in microstructure to be detected. Table 1 shows erosion test results on aluminas using both 160 μm silicon carbide (SiC) and 180 μm quartz (SiO_2) as the erodent, the SiO_2 being called a 'soft erodent' compared with the target alumina. It can be seen from this table that:

(i) The magnitude of erosion loss is greater in the case of SiC than SiO_2.
(ii) The relative performance of the aluminas is greater in the case of SiO_2 than SiC.
(iii) The two materials designated C1 and C2 are in fact the same materials, albeit from different batches, and normal mechanical testing such as hardness and bend strength, give similar results and yet the erosion test results are quite different. Microstructures of these two materials are shown in Fig. 4 and show the difference between these two 'identical' materials.

Theoretical models have been proposed to explain the mechanism of erosion in terms of mechanical properties but, when individual groups of materials are taken, the models fail to explain the differences.[2]

The theoretical models proposed[3,4] relate to properties of the target with properties of the erodent such that:

TABLE 1
Comparison of erosion losses for four grades of alumina

Material designation	Erosion loss ($cm^3\,t^{-1}$) ($45\ m\,s^{-1}$, 50° impact angle)	
	Silicon carbide	Quartz
A	180	10·1
B	240	21·9
C1	320	25·9
C2	525	42·2

$$W \propto V_o^a R^b \rho^c H^d K_{IC}^e$$

where W is the erosion volume loss per impact, V_o is the particle velocity, R and ρ are the particle radius and density, K_{IC} and H are the fracture toughness and hardness of the target and a, b, c, d and e are exponents.

The reason for the non-conformity to the models can be attributed to the scale of damage when compared with the scale of measurement of both hardness and fracture toughness. The scale of damage due to erosion is smaller than the grain size and therefore the wear resistance is due primarily to the strength of the grain boundary phase and the size of the grain when compared with the incoming particle. It is for this reason therefore that erosion testing is so specific as far as microstructure is concerned and why the models proposed cannot predict erosion loss based only on the macro properties of hardness and fracture toughness.

4. INSTALLATION CRITERIA AND METHODS

The selection of the grade of alumina is normally governed by the cost/performance ratio when taking into account wear resistance coupled with cost of installation, cost of scaffolding, cost of downtime, etc., but, as can be seen from earlier results, there is no one simple indicator of performance or combination of physical properties than can predict the comparative wear performance and certainly the use of properties such as alumina content or density in tender specifications should cease. Laboratory testing, or well controlled and monitored site testing, or preferably a combination of the two, are the only methods yet available to determine the suitability of an alumina. There is evidence, however, to suggest that static indentation tests on polished surfaces can semi-quantitatively give a guide to the suitability of an alumina to erosive attack.

Once the grade of alumina has been selected, there is then a series of choices which needs to be made dependent on the particular application before a lining is installed. These choices are simplified in Table 2. As mentioned in Section 3, alumina ceramics as used in this industry are normally manufactured by die pressing and slip casting and therefore the lining of components is carried out using either standard or special shapes. These tiles can vary in size from $25 \times 25 \times 3$ mm^3 to tiles of $250 \times 250 \times 50$ mm^3. The tile shape and size used will be dependent both on the life required of the lining and of the ease and cost of

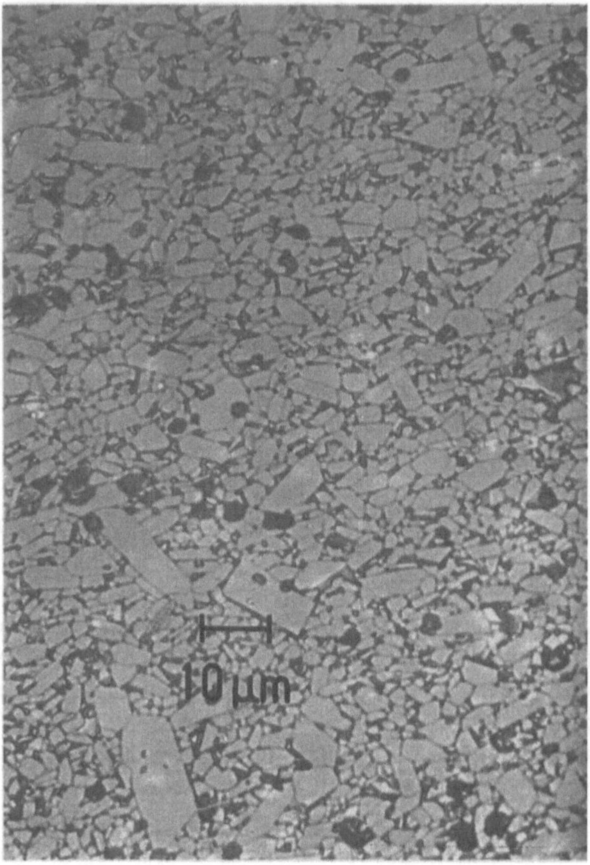

(a)

Fig. 4. Microstructure of two aluminas designated (a) C1 and (b) C2.

installation bearing in mind that steps against the direction of flow and gaps between tiles should be minimised.

Steps in the tile layout mean that a right angle impact is created which increases erosion rate and thus rapidly reduces the effective thickness of the lining and also creates local turbulences which can cause increases in impact angle local to the step. Gaps in the tile layout larger than the erosive particles themselves, particularly in the direction of flow, mean that the erosive particles can penetrate between the tiles causing erosion

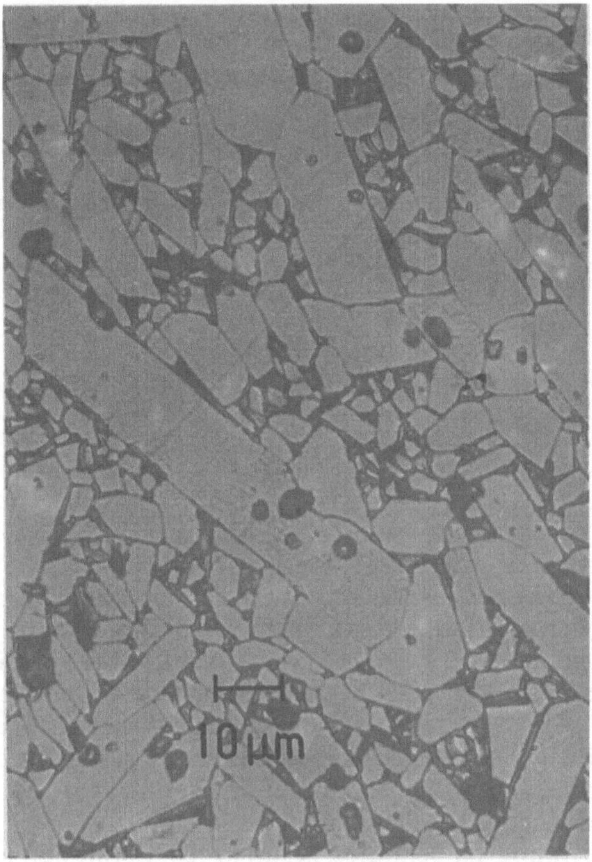

(b)

Fig. 4.—*contd.*

of the underlying substrate. Avoidance of this phenomenon will be discussed further in this section. If the direction of flow is known and velocities are high, then with typical pneumatically conveyed particles around 100–200 μm, it is impossible, except by using ground edge tiles, to reduce the gap between tiles to less than 300 μm. In these cases book-end or tongue-and-groove tiles can be used to increase the tortuosity for the travelling particles. Examples where book-end or tongue-and-groove tiles are required for this purpose include fan casings where

TABLE 2
Selection of processes for alumina as an erosion-resistant material

Material	Tile size	
	Tile shape	Book-end
		Plain
		Tongue-and-groove
Securement	Adhesive	Epoxy
		Silicone
		Silicate
	Weld-on	
	Mechanical interlock	
Tile layout	Brickwork	
	In-line	

velocities in excess of 80 m s^{-1} might be experienced. Tongue-and-groove and book-end tiles can also be used as a mechanical interlock to reduce the reliance on adhesives and this will be discussed further. Once the tile shape, quality, size and geometry have been decided for a particular application, then a method has to be devised for the securement of the tiles to the substrate. Factors that have to be taken account of include temperature of operation, dynamic/static forces on the tile, thermal conductivity of both material and substrate, condition of the substrate and environmental considerations.

At low operating temperatures, i.e. up to 150°C, two-part epoxy resin adhesives are extremely successful in that they are relatively easy to apply, have a pot life of 8 h under normal ambient temperatures and cure to full strength in 5 days. Properties of an epoxy resin which is finding increasing usage in the industry are shown in Table 3. At temperatures up to 230°C silicone adhesives can be used successfully, although their bond strength is relatively low when compared with the epoxy resins. At temperatures in excess of 230°C the choice of conventional adhesives is restricted to silicate cement type materials, which when heated become brittle and are liable to suffer from debonding due to mechanical shock. At high temperatures alternatives to adhesives are weld-on and/or bolt-on tiles and mechanical interlock. The main disadvantage of the weld-on/bolt-on lining in erosive situations is that the weld ferrules or bolts have to be protected from the

TABLE 3
Typical properties of flexibilised two-part epoxy resin[a]

Physical properties	
Pot life	4·5 h at 10°C 2·0 h at 20°C 1·3 h at 30°C
Typical substrates	Steel and concrete
Typical adherents	Ceramic, alumina, basalt, glass, etc.
Maximum service temperature	150°C continuous (occasional excursions to 180°C permissible for periods up to 24 h)
Mechanical properties	
Tensile strength (BS 2782, 320C)	15·0 MPa
Elongation at break (BS 2782, 320C)	8%
Elastic modulus (BS 2782, 320C)	1·35 GPa
Bend strength (BS 2782, 335A)	22·0 MPa
Vibration fatigue strength (ASTM D3166-73 (79))	6·0 MPa at 10 cycles

[a] Epron Tilebond (manufactured by BP Chemicals Ltd).

erodent. This is normally achieved with a ceramic plug which has to be glued in above the metal part, thus again placing reliance on adhesives at high temperature. Figure 5 shows an example of a tiled installation of a fan casing where weld-on tiles have been incorrectly used, not only because of the choice of weld-on tiles, but also because they have been installed in a brickwork pattern and normal impact has significantly reduced the life of the ceramic lining.

Much discussion and conjecture have centred around the method and layout of tiles relative to the flow direction. The main disadvantages put by both sides are that:

(a) In the case of brickwork pattern, a normal impact is created which will create local turbulences and will result in the erodent recirculating and eventually cutting through the underlying steelwork with little or no apparent wear of the ceramic.
(b) In the case of in-line joints, the erodent will, at the velocities concerned, gradually progress through the joints, particularly if the gaps between tiles are large when compared with the size of the erodent, and again penetrate the underlying steelwork.

Fig. 5. Alumina weld-on tiles on fan casing.

In complex dynamic situations, for example, fan blades, both the size of the tile and the properties of the adhesive have to be considered when taking into account the force on the fan blades and its deflection. It is also prudent in such circumstances to use a factor of safety of at least 6 since debonding of these tiles can result in damage to the fan casing lining and to the shaft bearing as a consequence of the fan becoming out of balance.

5. APPLICATIONS

Industrial applications where ceramics and particularly alumina ceramics have been used for the past decade to protect capital plant include classifiers, pneumatic conveying pipework, fan casings and, more recently, fan blades. Where alumina ceramics have been successful, and one class of material is not the panacea for all wear applications, then they have given significant cost effective advantages over conventional materials such as Nihard, mild steel and hardfacing. Although it is difficult or impossible to make general wear claims for

TABLE 4
Case history summary

	Case 1	Case 2
Application	Pneumatic conveying Pipework 75 mm ID 90° bends 600 mm CLR	Fan blades Diameter 2·2 m Rotational speed 980 rpm
Temperature	Ambient	80°C
Maximum particle size	4 000 μm	1 000 μm
Erosive material throughput	Aluminium oxide 5 t h^{-1}	Iron oxide —
Tile shape	89 × 89 × 13 mm^3 + special cast shape	50 × 50 × 4 mm^3
Adhesive	Two-part epoxy resin and mechanical	Two-part epoxy resin

materials, two specific case histories are cited which have been in service for more than 2 years and have been shown by quite simple costing to be cost effective, see Table 4. Both these applications, one in a sensitive environmental area and the other in a high technology demanding dynamic situation show cost effectiveness even when the cost of plant/production shutdown, the cost of scaffolding, the cost of standby plant, etc., are not taken account of.

In the case of the aluminium oxide transport it was predicted that 6 mm mild steel bends would last, on the basis of pneumatic conveying trials, for 6 h only, i.e. a throughput of 30 t, whereas in practice the 13 mm alumina tiles have been in service for 4 years, i.e. a throughput of 87 000 t. In this particular application the bends are some 13 m in the air, necessitating scaffolding for replacement and lost production during replacement and repair.

In the case of the fan blades 4 mm thick weld deposit tungsten carbide had been used previously and this had lasted between 3 and 6 months; application of the ceramic tiles has increased the life to date to 30 months. In this application where the fan is part of a production process, opening up of the fan is a time consuming and expensive process. The application of ceramic tiles has not only reduced this maintenance cost and given continuous production but has also increased the life of the shaft bearings and reduced vibration caused by being out of balance.

As experience is gained by operators and as engineers become aware of other classes of materials in addition to metals then new uses will be found for ceramics. The growth of adhesion technology and the continued development of alumina ceramics to increase the toughness will also result in the increasing use of these materials to protect and enhance the life of metal components.

REFERENCES

1. Morgan Roctec — Internal Marketing Report.
2. Dimond, C. R., The relationship between the erosion performance of oxide ceramics and their microstructure and physical properties. *Proceedings of Eurotrib '85*, 1/6–6/6. Elsevier Science Publishers BV, Amsterdam.
3. Oh, H. L., Oh, D. L., Vaidyanathan, S. and Finnie, I., The shaping of brittle solids by erosion and ultrasonic cutting. Special Publication 348, National Bureau of Standards, Washington DC, 1972.
4. Evans, A. G., Gulden, M. E. and Rosenblatt, M., Impact damage in brittle materials in the elastic–plastic response regime. *Proc. R. Soc. London,* **A361** (1706) (1978) 343–65.

7

Silicon Nitride Products Development for Automotive Engines

KEIJI MATSUHIRO AND MINORU MATSUI

NGK Insulators Ltd, Nagoya, Japan

1. INTRODUCTION

NGK started to tackle the development of structural ceramics for heat engines in the early 1970s. The primary intention of the development was to challenge and innovate the new ceramic technology applicable to the wide range of high stress and high temperature structural components used for gas turbine and reciprocation engines. Pioneer work had already been carried out in the USA, Europe and Japan by academic, government and industrial workers.

Since then, studies of materials, design methodology, and engineering and quality assurance methods have continued, aimed at the mass production of components. Various kinds of key technologies born from this development have been adapted, not only to the high-tech fields, but also to pre-existing conventional products.

The development of structural ceramics is primarily addressed, as is the integration of fundamental technology, by most of the Japanese ceramic manufacturers, such as NGK.

2. DEVELOPMENT HISTORY

A brief summary of turbocharger rotor and pressure wave supercharger rotor development work will be given.

2.1 Turbocharger Rotor

Major tasks for the development of the turbocharger rotor were as follows:

1. development of high strength pressureless sintered silicon nitride materials;
2. process control of injection moulding;
3. engineering of ceramic–metal joints.

2.1.1 Material

Sintered silicon nitride materials have the most promising advantages compared to other structural ceramics. The well known interlocked, and elongated, silicon nitride grain microstructure can provide higher strength and higher toughness than that of other equiaxed grain ceramics. However, this characteristic simultaneously gives lower sinterability, and the tendency to retain the intergranular glassy phase. In order to overcome these problems, raw powders, sintering additive systems, densification process control, microstructure optimization, and mechanical property characterization, have been investigated.

Table 1 shows the properties of one of the latest production silicon

TABLE 1
Material properties of SN-84

Density (Mg m^{-3})	3·24
Bend strength (four-point) (MPa)	
RT	860
800°C	800
1 000°C	760
1 200°C	780
1 400°C	150
Young's modulus, RT (GPa)	280
Poisson's ratio, RT	0·25
Fracture toughness, RT (MP $m^{1/2}$)	
RT	6·7
1 000°C	6·1
1 200°C	7·0
Hardness, RT (Knoop 300 g load) (GPa)	16·0
Thermal expansion coefficient (40–1 000°C) (MK^{-1})	3·8
Thermal conductivity (W m^{-1} K^{-1})	
RT	37
1 000°C	16

nitride materials (SN-84). The properties listed are collected from a cut-out bend bar of the production turbocharger rotor, produced by injection moulding and nitrogen atmosphere sintering. This material consists of β-Si_3N_4 grains, crystalline intergranular phases and a small amount of glassy phase. A typical microstructure is shown in Fig. 1. The lattice image of the high resolution transmission electron micrograph indicates that the β-Si_3N_4 grains, and the secondary crystalline intergranular phases identified as J-phase and H-phase, are separated by an approximately 5 nm thick glassy phase. Chemical composition and grain morphology control throughout the densification process are the keys to achieving the good in service durability and reliability. This material can maintain good strength up to 1200°C with controlled time-dependent stability. From laboratory scale material research up to the volume production development, researchers and product engineers carried out a significant amount of collaborative work.

2.1.2 Injection moulding process

The injection moulding process was the most challenging task when the development was started. Very few studies had been done at that time, and, in general, injection moulding as a ceramic forming technique was quite a new process. Problems of binder system selection, moulding die design, and mass production de-waxing without leaving critical defect, had to be solved.

Typical defect locations are illustrated in Fig. 2. 'Short shot' means insufficient material filling at the shape narrow end. Cracks frequently occur in the transition area of the shape discontinuity. Weldline is introduced by a lower compaction uniformity of the prepared noodle material. Pores are caused by inadequate mixing of the raw powder and wax systems. Unmelted wax, or local explosively evaporated wax, forms this type of low density pocket inside the compacted body. All these types of defects should be minimized below the critical level. Additional difficulties with the injection moulding process are illustrated in Fig. 3 which shows the general relationships between powder particle size, sintered body strength, and shaped body dewaxability. The finer powders are better for strength, but poor for dewaxability. This trade-off has to be solved in meeting the need for required material strength, without sacrifice of an economical dewaxing cycle. Research to provide understanding of powder–wax mixture characteristics, and wax migration path control mechanisms, was the critical task in this respect. Large-scale rescue work was organized, to classify the parameters influencing individual processes and to select the appropriate solutions.

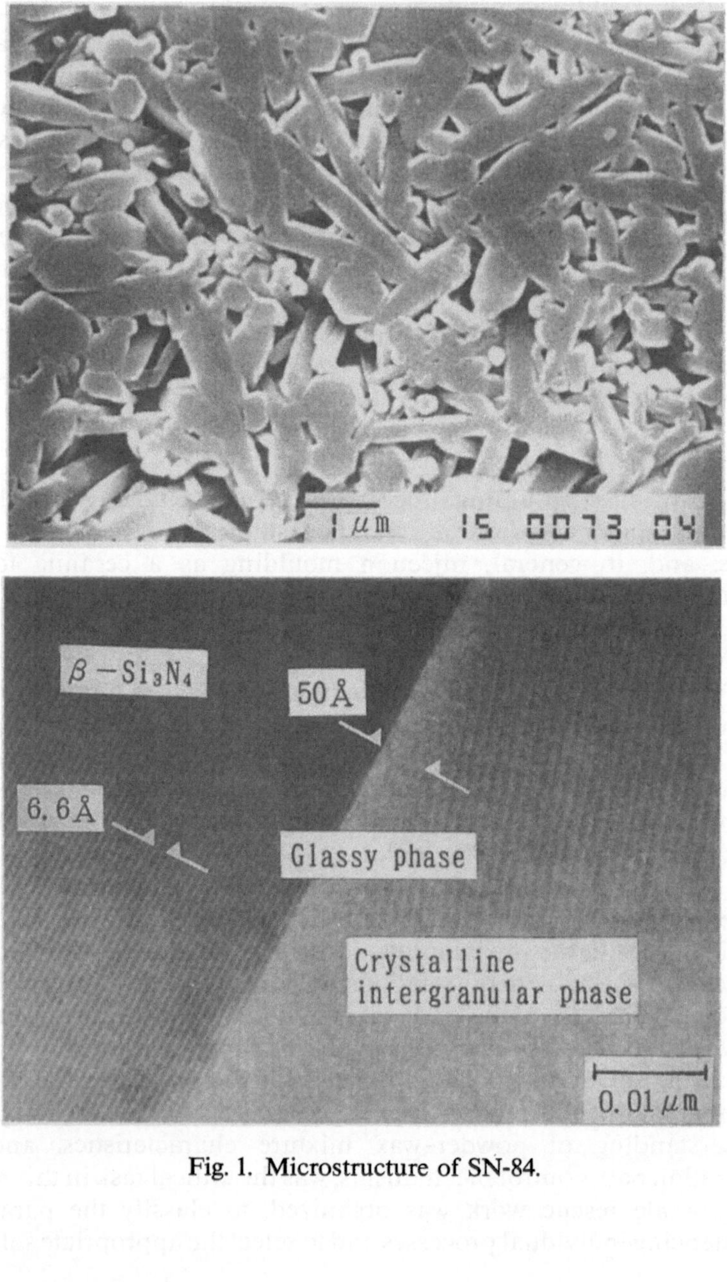

Fig. 1. Microstructure of SN-84.

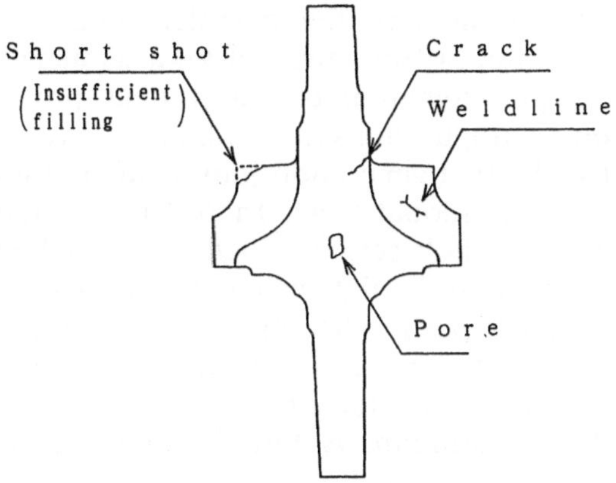

Fig. 2. Typical defect location of injection mould turbocharger rotor.

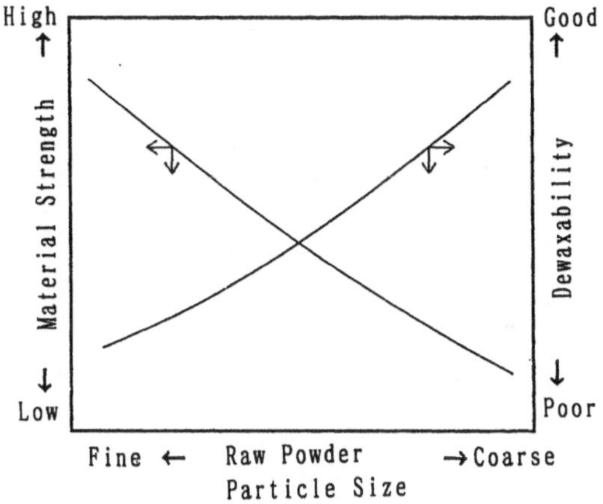

Fig. 3. Particle size, dewaxability and material strength characteristics of injection moulding products.

2.1.3 Engineering of ceramic–metal joints

Suitable mass production methods for joining the ceramic rotor to the metal shaft were also not available for these types of severely stressed components. Thermal shrink fitting and braze bonding were selected as potential methods.[1-3] Competitive technical studies were set up on the various types of configuration. Because of the process controllability and simplicity, the thermal shrink fitting method was selected by NGK.

However, in order to achieve the complete quality assurance, a significant amount of effort was directed towards metallurgy, design, rig-testing and engine dynamometer and field proof-testing. The currently used joint configurations for a commercial vehicle are shown in Fig. 4. Type I is the first generation joint configuration. The entire metal shaft is a conventional steel. In order to achieve adequate gripping force over the whole ceramic–metal joint the low temperature portion was designed with the tighter interfacial force, and combined with a high temperature portion having a lower interfacial force. This idea is one of the solutions using thermal expansion mismatch materials such as silicon nitride and steel. Type II is the second generation joint configuration. A low thermal expansion alloy is

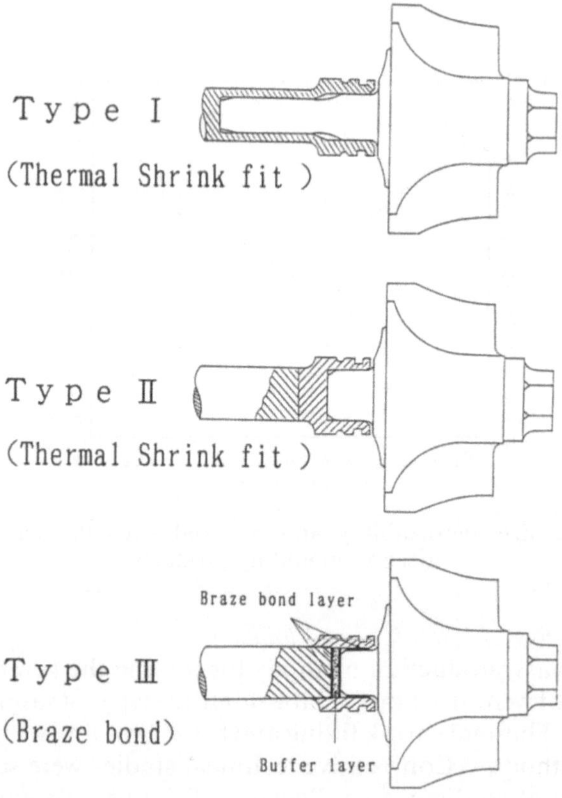

Fig. 4. Currently used ceramic rotor-metal shaft joint configurations.

selected as the rotor gripping link between the steel and the silicon nitride. With this configuration, a relatively expensive low thermal expansion alloy cup can deal with the reduction in the machined area of the silicon nitride rotor. Type III is also available for the ceramic–metal joint. The thermal mismatch absorption buffer layer between the steel shaft and the silicon nitride, together with a braze bond alloy, was examined. Investigations into suitable alloy systems were also carried out.

The combined efforts of ceramic, metal, design and engineering studies were required in order to achieve adequate durability and reliability.[4]

2.2 Pressure Wave Supercharger Rotor

Work similar to that on turbocharger development has been carried out. A new approach in this development was to employ the conventional extrusion forming process with the high-tech structural ceramics.[5] A unique sophisticated supercharging concept innovated by BBC (a Swiss based engineering company, now renamed ABB Turbo) was coupled with 'state-of-the-art' ceramic components.[6,7] An extruded rotor is shown in Fig. 5. The three-layered individual through-hole geometry is carefully designed to pump up the intake air by the exhaust gas pressure wave, with a significantly higher efficiency, compared to other charge–air systems, such as the turbocharger or mechanical supercharger.

Materials properties requirements were set up initially to assure attainment of the rotor duty specification. Tables 2 and 3 list in detail a number of the duty specifications and properties. Engineers from both BBC and NGK had talked seriously to decide these values. At the very beginning the materials properties had not been verified completely to assure the required duty specifications. Therefore, material property measurements to provide the data base for the design methodology had first to be carried out. Most of the time-dependent type characteristics were examined to assure the entire operating life. For some of these measurements the test method itself was quite new for ceramics. In order to assure the optimized material properties of the silicon nitride (SN-75), tests exclusively developed for this application were used to assess its characteristics, as shown in Figs 6, 7, 8, 9 and 10. For example in Fig. 9, the silicon nitride material developed showed a similar type of Goodman fatigue phenomena to metals. Figure 10 means that the calculated rotor failure rate at 5000 h satisfies the required number of $<10^{-5}$.

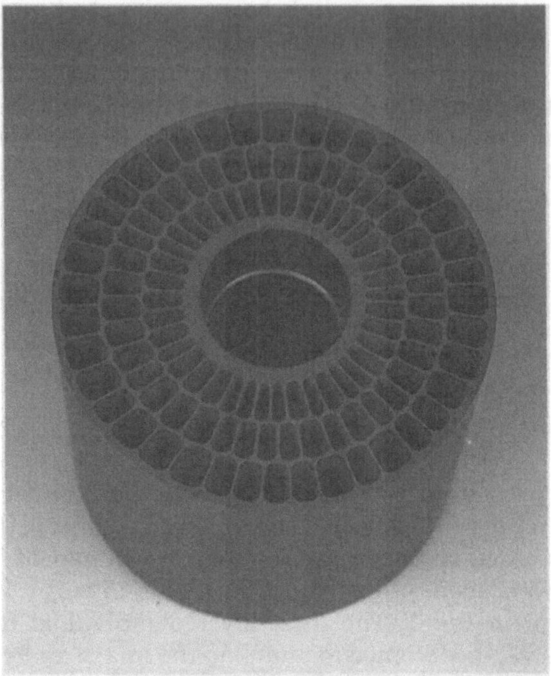

Fig. 5. Extruded SN-75 rotor.

Design methodology for structural ceramics was associated with simultaneous material, fabrication, and engineering development.[8–13]

3. OVERVIEW OF THE CERAMIC INDUSTRY MOVEMENT

In Japan, structural ceramics have already been launched into the commercial market. The development history experienced by NGK is only one example of the efforts being made in Japan. Each company is aiming to expand activities based on individual business plans. Automotive engines, gas turbines, precision machining, chemical equipment, cutting tools, metal handling equipment, refractory equipment, sporting goods and so on, are currently identified as potential markets. Daily accumulation of technological successes will enable the fields of application of structural ceramics to expand.

TABLE 2
Pressure wave supercharger rotor duty specification

Nominal circumferential velocity (m s^{-1})	80
Temperatures	
Gas max. (°C)	850
Rotor max. (°C)	650
Gradient radial (°C)	75
Gradient axial (°C)	500
Thermal expansion (MK^{-1}) (axial clearance, leakage)	<4
Thermal conductivity (W m^{-1} K^{-1}) (stresses)	>21
Density (Mg m^{-3}) (weight, inertia, stresses)	<3·50
Atmosphere	Diesel exhaust and condensate
Life for passenger car applications (h)	>3 000
Failure rate	$<10^{-5}$
Factor of safety proof test/service	>2

TABLE 3
Material properties of SN-75

Density (Mg m^{-3})	3·20
Bend strength (four-point) MPa	
RT	470
800°C	440
Young's modulus, RT (GPa)	275
Poisson's ratio, RT	0·27
Fracture toughness, RT (MPa $m^{1/2}$)	5·9
Cyclic fatigue strength (MPa)	
20°C 10^8 cycles	>250
650°C 10^8 cycles	>250
Static fatigue strength (MPa) 1 000 h at 750°C	>250
Thermal shock resistance ΔT (°C)	550
Thermal conductivity ($Wm^{-1}K^{-1}$)	29
Thermal expansion coefficient (MK^{-1}) (40–600°C)	3·1
Resistance to oxidation (g m^{-2}) Weight gain 1 000 h at 750°C	1·0
Weibull modulus for four-point bend specimens	14

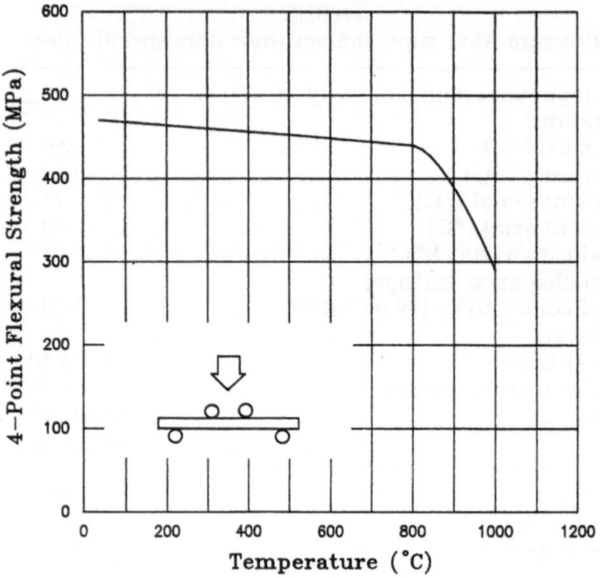

Fig. 6. Four point flexural bend strength of SN-75.

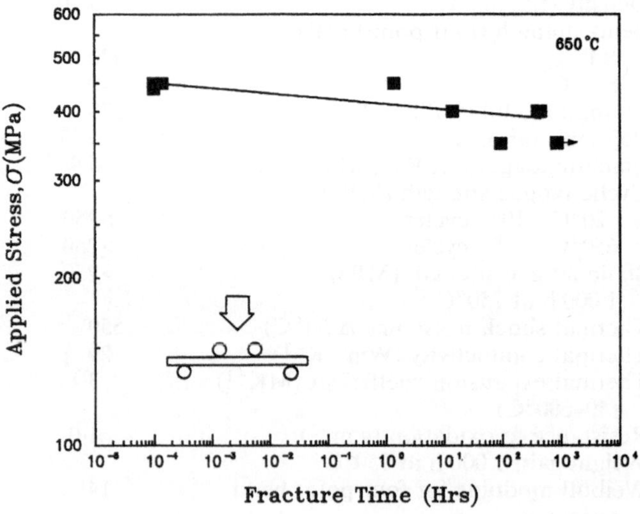

Fig. 7. Static fatigue of SN-75.

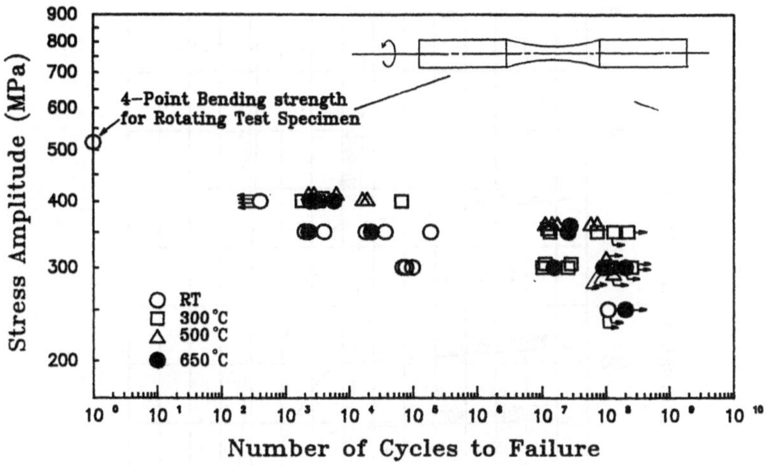

Fig. 8. Cyclic fatigue of rotating-bend test of SN-75.

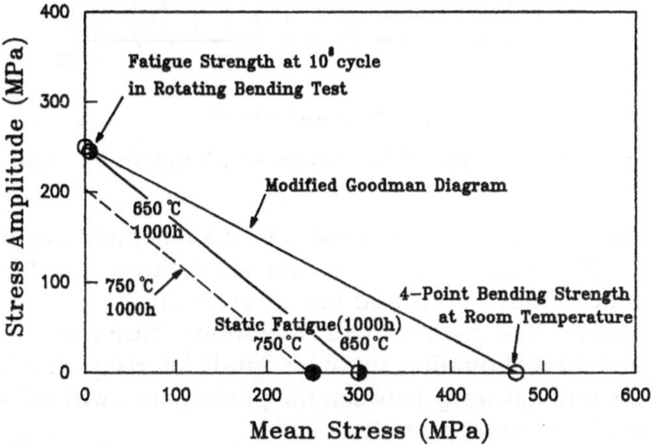

Fig. 9. Goodman fatigue diagram of SN-75.

4. CONCLUSIONS

The structural ceramics have potential advantages in several specific applications. However, ceramics are still rather new and developing industrial materials. Property databases, design methodology, quality assurance, non-destructive inspection, and standardization of property

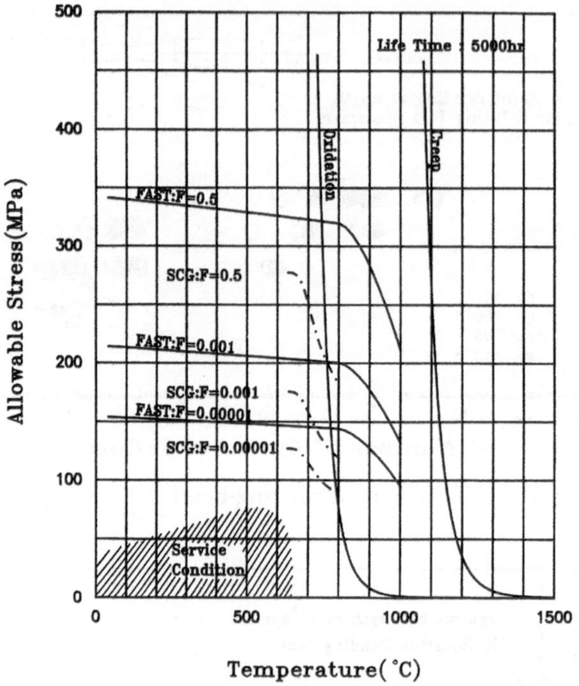

Fig. 10. Failure probability prediction for the SN-75 rotor.

measurements, are not yet completed. The ceramic manufacturers are planning to invest, in order to settle these tasks, through collaboration with academic organizations and the government.

Each country has its own considerations based on historical background and nationality, in these kinds of risky development. Competition, and harmony between the participants, will enhance the growth rate of structural ceramics.

ACKNOWLEDGEMENTS

The authors wish to thank Mr A. Mayer of ABB Turbo Systems Ltd, for giving us the ready permission to use data. The authors also wish to thank all the people in NGK for their cooperation and assistance in the preparation of this paper.

REFERENCES

1. Matoba, K., Katayama, K., Kawamura, M. and Mizuno, T., *The Development of Second Generation Ceramic Turbocharger Rotor — Further Improvements in Reliability*, SAE Technical Paper Series 880702, Detroit, 1988.
2. Kuwabara, Y., 'Light-weight, heat-resistant materials supporting high-performance Nissan car', *Nikkei New Materials,* **9/18** (1989) 64–71.
3. Okazaki, Y., The development of ceramic turbocharger rotor. *FC Report,* **3**(8) (1985) 16–23.
4. Nishiguchi, F. and Matsuo, Y., Second Generation Ceramic Turbocharger Rotor. *FC Report,* **6**(9) (1988) 361–3.
5. Mayer, A., Oda, I., Kato, K., Haase, W. and Fried, R., *Extruded Ceramic — A New Technology for the Comprex-Roter*, SAE Technical Paper Series 890425, Detroit, 1989.
6. Hiereth, H., *Car Tests with a Free Running Pressure-Wave Charger — A Study for an Advanced Supercharging System*, SAE Technical Paper Series 890453, Detroit, 1989.
7. Zehnder, G., Mayer, A. and Matthews, L., *The Free Running Comprex*, SAE Technical Paper Series 890452, Detroit, 1989.
8. Soma, T., Matsui, M. and Oda, I., Tensile strength of a sintered silicon nitride. In *Non-Oxide Technical and Engineering Ceramic*, ed. S. Hampshire, Elsevier, London, 1986, pp. 361–74.
9. Oda, I., Matsui, M., Soma, T., Masuda, M. and Yamada, N., Fracture behavior of sintered silicon nitride under multiaxial stress state. *J. Ceramics Soc. Japan,* **96**(5) (1988) 539–45.
10. Masuda, M., Soma, T., Matsui, M. and Oda, I., Fatigue of ceramics (Part 1). Fatigue behavior of sintered Si_3N_4 under tension-compression cyclic stress. *J. Ceramics Soc. Japan,* **96**(3) (1988) 277–283.
11. Masuda, M., Yamada, N., Soma, T., Matsui, M. and Oda, I., Fatigue of ceramics (Part 2). Cyclic fatigue properties of sintered Si_3N_4 at room temperature. *J. Ceramics Soc. Japan,* **97**(5) (1989) 520–24.
12. Masuda, M., Soma, T., Matsui, M. and Oda, I., Fatigue of ceramic (Part 3). Cyclic fatigue behavior of sintered Si_3N_4 at high temperature. *J. Ceramics Soc. Japan,* **97**(6) (1989) 612–18.
13. Soma, T., Ishida, Y., Matsui, M. and Oda, I., Ceramic component design for assuring long-term durability. *Advanced Ceram. Material,* **2**(4) (1987) 809–12.

8

Revolution in Ceramic Processing

K Kendall

ICI, PO Box 11, Runcorn, Cheshire, WA7 4QE, UK

1. INTRODUCTION

Ceramic manufacturing processes are advancing vigorously at the present time. Although most engineering ceramics are still made by the traditional powder pressing, slip casting or extrusion routes, followed by green machining, firing and diamond machining,[1] there is a growing use of novel processes in the technical areas of engineering ceramics.[2] Vapour deposition, sol–gel processing, reaction forming and colloidal processing immediately spring to mind as examples. These areas may be relatively small in commercial terms at the moment, but they represent a revolutionary change in process thinking.

The reasons for this revolution are numerous: there is the driving force of the electronic ceramic industry which has devised new gas phase or fluid methods for depositing ceramic layers; there is the move towards chemical precursors which can guarantee purity and performance; there is the continuing discovery of new materials or composite structures which demand novel processing solutions; there is the slow but steady shift away from machining towards moulding, leaning heavily on the plastics injection moulding experience; but most of all, there is the realisation that the properties of ceramic products can be much improved through processes which deliver better microstructure (Fig. 1).

The purpose of this paper is to summarise and collate the progress of these new ideas. First, the negative features of traditional ceramic processing are described. Then some of the new processes are reviewed,

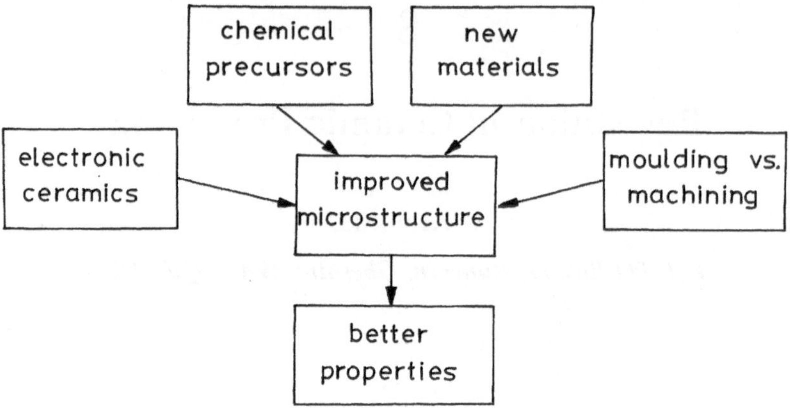

Fig. 1. Driving forces for improvements in ceramic processing.

bearing in mind that certain innovations represent incremental changes in traditional methods whereas other advances are truly fundamental. Finally, two detailed case histories of radical inventions are considered: sol–gel alumina fibre which emerged as a product in 1974;[3] and colloidal processing which was first embodied in cement products in 1988,[4] is now used to make ceramic superconductors,[5] and will soon appear in engineering parts.

1. PROBLEMS WITH TRADITIONAL PROCESSING

Ceramics are not easy to process. The better the ceramic in terms of its resistance to heat, to chemical attack and to plastic flow, the more difficult it is to manufacture the product. The paradox is that the most perfect engineering ceramic would be unprocessable. So the history of ceramic advance may be regarded as a sequence of processing inventions which make ceramic production possible.

In ancient times, the best ceramics were found as natural deposits, then cleaved or machined to shape: diamonds, aluminas, silicas and other gemstones. Cutting was the original process for ceramic production, and still represents the costliest operation in much ceramic manufacture, besides damaging the microstructure as a result of surface cracking.[6]

The invention of sintering in Babylonian times was a breakthrough.[7]

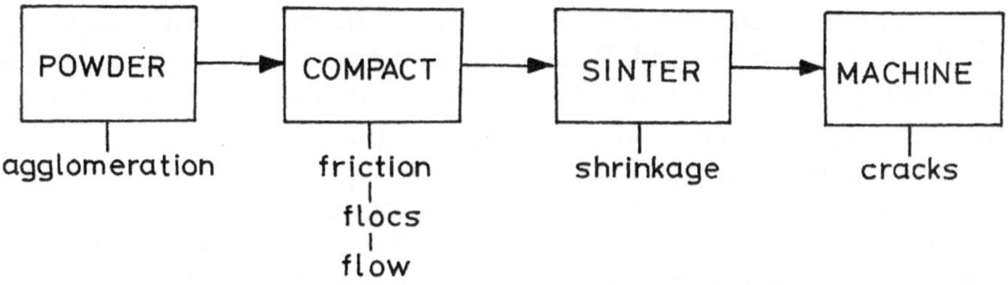

Fig. 2. Problems of a conventional ceramic process.

From a powdery starting material, a shape could be moulded and fired in a furnace at modest temperature to obtain near full density. All materials are known to be sinterable below 2000 K given a fine compacted powder. The major drawback is the shrinkage that occurs during sintering. Final machining is almost always necessary as a result of this firing shrinkage. The other difficulties are the supply of sufficiently fine powder, and the compaction of the powder in the mould (Fig. 2).

Traditionally, powder compaction is by dry pressing of powder, either with rigid dies or with fluid pressure applied to a rubber membrane, to mould small, simple-shaped articles.[8] However, fine dry powders do not flow readily into moulds and do not behave like fluids under compaction because friction prevents easy movement of grains. Consequently there are density variations through the moulding, and agglomerates remain as defects in the final piece.

In slip casting, the powder is dispersed into a liquid to form a slurry which is poured into moulds which may be porous so that the liquid drains out to leave a solid deposit of ceramic particles on the mould surface. This is a cheap process and one which is suitable for making large and complex items such as concrete buildings and sanitary ware. But the properties of the final product are rarely better than those of pressed materials because flocculation occurs as a result of van der Waals forces between particles, causing aggregation, shrinkage and cracking.

Extrusion of plastic compositions can also be used to make large pipes, rods and sheets. Originally, such extrusion was based on the flow properties of clays which exhibit Bingham rheology, that is an ability to support load below a certain pressure and to flow above that pressure. Nowadays, a wide range of polymer flow aids is available to allow extrusion and injection moulding of any ceramic powder, in both aqueous and solvent systems. The principal difficulty is dispersing the powder in the carrier fluid. Because agglomerates survive this mixing process, the final ceramic is defective and offers no property advantage over pressing or casting, while drying and polymer burn-out remain expensive additional problems.

In summary, conventional powder compaction and sintering processes produce useful products but suffer from powder handling problems, excessive shrinkage and poor microstructures. To overcome these difficulties, new processes have been devised.

3. NEW PROCESSES

New processes can be of two kinds, the sudden increment in existing technology which pushes the products into new areas, and the fundamental new concept which develops over a relatively long period.

It would be neat if a completely new material came together with a totally novel process to create a radical change in products. However, this conjunction of two rare discoveries seldom arises. For example, superconducting ceramics were initially fabricated by the traditional calcining, grinding, pressing, sintering route. Partially stabilised zirconia is generally made by pressing spray-dried powders followed by sintering and machining (Fig. 3). However, once established, these materials may be fabricated by novel processes to obtain optimum products. For instance, superconducting films may be made by laser ablation; partially stabilised zirconia may be made by colloidal processing.

One can identify several impressive incremental developments in the traditional ceramic processes of pressing, casting and extrusion. These may be taken together and classified as engineering improvements (Fig. 4).

Pressing has been enhanced by hot isostatic pressing (HIPing)[9] which uses gas as the compression medium at high temperatures. Residual pores and defects are compacted to improve the structure of the product. Thus, the strength and reliability of the product is raised. Another technique which has been demonstrated recently is superplastic forming of zirconia ceramic at elevated temperature.[10] Complex shapes of dense ceramic can be forged from presintered material by pressing in shaped dies.

Casting has been improved in the pressure casting system[11] which

Fig. 3. Nilcra stabilised zirconia product; bend strength of 716 MPa, Weibull modulus of 30, and toughness of 15 MPa $m^{1/2}$ are typical.

fills the mould under pressure, thereby removing bubbles. Plastic moulds are now commonly used instead of the traditional plaster. Tape casting,[12] developed in the electronic ceramic industry, has been used to make complex composite structures. The ceramic powder is dispersed in a polymer/solvent mix, doctor bladed onto a moving plastic support film, and oven dried. The resulting smooth, thin ceramic tape can be peeled from the support, handled just like a polymer film, and assembled into a multilayer structure before burn-out and firing. Casting has also been developed to make ceramic foams by coating a ceramic slurry onto a plastic foam precursor which is later removed in the firing procedure.

Extrusion has benefited from the invention of the matrix die which

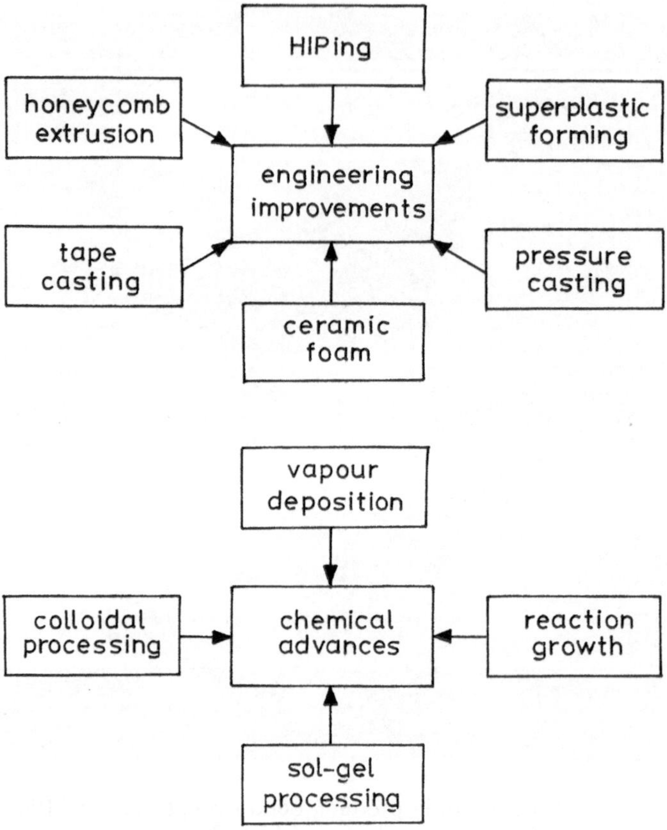

Fig. 4. Summary of new processing options under the headings of engineering improvements and chemical advances.

enables a ceramic honeycomb of staggering complexity to be simply extruded from a plastic mix. This is surely one of the most elegant engineering devices of recent years.[13]

4. RADICAL NEW PROCESSES

None of the above improvements can be considered to be fundamental because none has altered the micro or nanostructure of the ceramic product. Only the outer appearance of the product is modified. In this section we consider the new processes which substantially alter the way in which the ceramic is formed by manipulating atomic or molecular forces. These developments may therefore be grouped together under the heading of chemical processes (Fig. 4).

One process which compares in significance with the invention of sintering is vapour phase deposition.[14] In the first instance this technique was developed by the electronics community to grow thin films by physical vapour deposition (PVD; sputtering, evaporation, ablation, etc.) or by chemical vapour deposition (CVD; reaction of gas to deposit ceramic layers). Because of low growth rates and the build-up of residual stresses, vapour phase deposition is ideal for films, fibres, or composite binding, using SiC, TiN, BN, C, etc. There is the great benefit of zero shrinkage in this process. But the most significant feature of such materials is their fine microstructure and high strength. This is illustrated in Fig. 5 which contrasts the strength and Weibull modulus of silicon carbide made by powder pressing and by CVD. The SiC made by vapour deposition is superior by an order of magnitude because of its finer structure.

Another method for production of novel microstructure is reaction growth of ceramic from either the solid or liquid phase. For example, the electrolytic anodising process used for making alumina membranes[15] is a fascinating invention. At high fields, the anodic layer grows with large pores but as the field is reduced the pores grow smaller and eventually reach nanometre dimensions before the membrane floats from its metal substrate. In the Lanxide process[16] molten aluminium is oxidised on its surface at high temperature to grow a ceramic layer of considerable

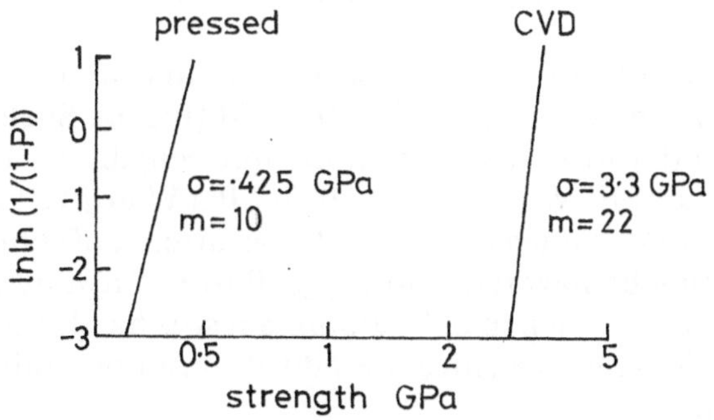

Fig. 5. Contrast between strength and reliability of SiC made by powder method and by CVD.

thickness. The final product is a composite structure which retains some metal content.

Sol–gel processing,[17] in which a ceramic is made directly from a chemical precursor, has also made a significant impact over the past few years. This method is most used to make pure powders, fibres and films, starting from nitrates, chlorides, metal organic compounds such as oxalates, citrates and ethoxides, or from organic polymers. Perhaps the greatest revolution in ceramics is that most technical powders are now made through chemical routes. Sol–gel processing is used commercially to make packings for chromatography, films for anti-reflection coatings, and fibres.

Colloidal processing[18] is a relatively new development based on the idea that microstructure of ceramics made by sintering powders is largely governed by van der Waals forces, the weak flocculation forces which cause agglomeration. By altering these molecular forces, significant improvements in properties such as strength and reliability have been demonstrated. Examples are the development of macro-defect free cement about ten years ago,[19] and extrusion of super-conducting wires.[5]

5. CASE HISTORY: SOL–GEL ALUMINA FIBRES

The first commercial product based on sol–gel technology was 'Saffil' alumina fibre made by spinning aluminium oxychloride solution through spinneret holes, drawing the fibres down in a dry air stream.[20] The dried threads are collected and heat treated to remove trace water, hydrogen chloride and organic matter before sintering to the correct ceramic phase.

The main benefit of the process is that fine alumina fibres of controlled diameter, near 3 μm, are obtained (Fig. 6). Such fine fibres allow the material to be used in short fibre lengths as an excellent thermal insulator (0·1 W m^{-1} K^{-1} at 200°C, 0·5 W m^{-1} K^{-1} at 1600°C). The control of fibre diameter prevents the occurrence of submicrometre fibres which may be hazardous to lungs. Purity is readily adjusted in this chemical process such that the alumina resists harsh environments. In particular, dopants like silica are added to inhibit grain growth at high temperature.

Because the microstructure is fine and the surfaces are smooth, the fibres are mechanically strong with a tensile strength above 1 GPa. This

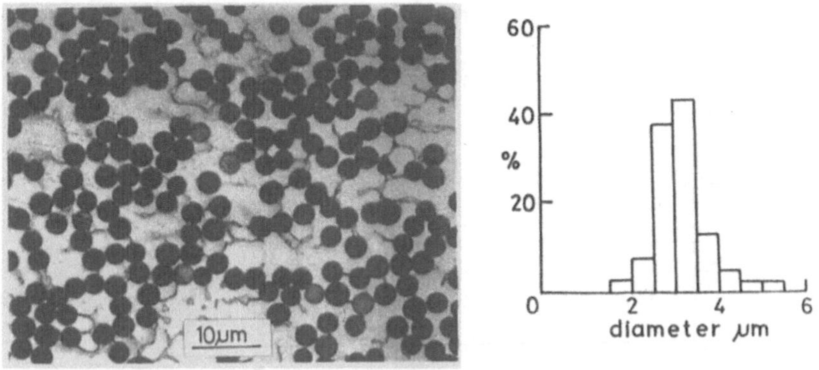

Fig. 6. Micrograph of Saffil alumina fibre showing the control of fibre diameter.

makes them useful for reinforcement of aluminium alloys. Longer fibres are now developed for use in this application.

6. CASE HISTORY: COLLOIDAL PROCESSING

Ceramics made by conventional powder forming techniques such as pressing, slip-casting or extrusion are always much weaker than products made by sol–gel or vapour deposition methods. For example, pressed alumina is typically 500 MPa in bend strength whereas sol–gel alumina fibre is around 2000 MPa. It has been demonstrated that this discrepancy is caused by the action of van der Waals forces in powder systems. These forces agglomerate fine particles into large clumps which act as defects in the final product. During slip-casting, the clumps are formed in the drying step; in powder pressing and extrusion the lumps are present in the original powder and are too strong to be broken down during the compaction stage.

Two methods have been outlined for removal of such weakening agglomerates to improve the microstructure and properties of ceramics made from powders. In the first approach, brute force is applied to a plastic mix of ceramic powder and polymer solution by extruding through a narrow orifice. The second method involves the production of a fully dispersed submicrometre suspension, preventing the coagulation into clumps by changing the drying procedure. In this way, much improved strength and reliability have been obtained for titania,

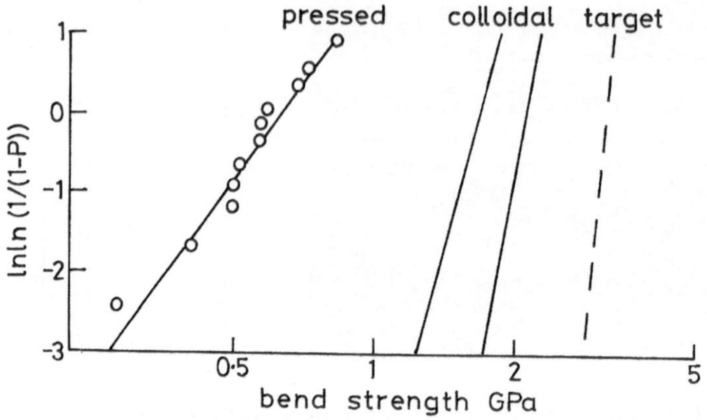

Fig. 7. Improvement in zirconia mechanical properties after colloidal processing.

alumina, silica, silicon carbide and zirconia ceramics made by the sintering route.[21] Figure 7 shows improvements for 3% yttria-stabilised zirconia. The left hand curve shows results for die pressed powder. The right hand curve indicates the target properties achievable if all agglomeration was prevented. The intermediate lines show the results obtained by colloidal processing the same powder at two extrusion pressures, illustrating the gain in strength and reliability.

7. CONCLUSIONS

The established ceramic processes of pressing, casting, sintering and machining will not disappear. They are being developed incrementally to suit new materials such as zirconia, silicon carbide and silicon nitride. But radically new processes such as vapour deposition, reaction forming, sol–gel and colloidal processing, which are still minor in commercial terms right now, will undoubtedly advance further as the drive to achieve improved microstructure and properties continues.

REFERENCES

1. Mangels, J. A. (ed.), *Forming of ceramics*, American Ceramic Society, Columbus, OH, 1984.

2. Alford, N. McN., Birchall, J. D. and Kendall, K., Engineering ceramics - the process problem, *Mater. Sci and Technol.,* **2** (1986) 329–36.
3. Birchall, J. D., Bradbury, J. A. A. and Dinwoodie, J., Alumina fibres — preparation, properties and applications, *Handbook of composites, Vol. 1,* W. Watt and B. V. Perov, (eds), Elsevier, Amsterdam, 1985, pp. 115–54.
4. Kendall, K., Birchall, J. D. and Howard, A. J., The relation between porosity, microstructure and strength, and the approach to advanced cement-based materials, *Phil. Trans. R. Soc. Lond.,* **A310** (1983) 139–53.
5. Alford, N. McN., Birchall, J. D., Clegg, W. J., Harmer, M. A. and Kendall, K., Physical and mechanical properties of $YBa_2Cu_3O_{7-d}$ Superconductors, *J. Mater. Sci.,* **23** (1988) 761–8.
6. Kishimoto, A., Koumoto, K., and Yanagida, H., Comparison of mechanical and dielectric strength distributions for variously surface finished titanium dioxide ceramics, *J. Am. Ceram. Soc.,* **72** (1989) 1373–6.
7. Kingery, W. D., Bowen, H. K. and Uhlmann, D. R., *Introduction to ceramics*, John Wiley, New York, 1976, pp. 469–513.
8. Norton, F. H., *Fine ceramics*, McGraw-Hill, New York, 1970, Chapter 10, pp. 130–56.
9. Larker, H. T., Dense ceramic parts hot pressed to shape by HIP. In *Emergent process methods for high technology ceramics, Vol. 17,* R. F. Davis, H. Palmour and R. L. Porter (ed.), Plenum Press, New York, 1984, p. 571.
10. Wakai, F., Kadama, Y., Sakaguchi, S., Murayama, M., Izaki, K. and Niihara, K., A superplastic covalent crystal composite, *Nature,* **344** (1990) 421–3.
11. Gebruder Netzsch Machinenfabrik GmbH, Technical Information Bulletin GK012, D8672 Selb, Bavaria, FRG, 1985.
12. Mistler, R. E., *Am. Ceram. Soc. Bull.,* **69** (1990) 1022–6.
13. Benbow, J. J. and Lord, L. W., Catalyst support, US Patent 3824196, 1974.
14. Spear, K. E., Diamond — ceramic coating of the future, *J. Am. Ceram. Soc.,* **72** (1989) 171–91.
15. Furneaux, R. C., Rigby, W. R. and Davidson, A. P., The formation of controlled-porosity membranes from anodically oxidised aluminium. *Nature,* **337** (1989) 147–9.
16. Lanxide Corp., One Tralee Industrial Park, Newark DE19711-54444, USA.
17. Segal, D. L. and Woodhead, J. L., Novel developments in gel processing. *Brit. Ceram. Proc.,* **38** (1986) 245–50.
18. Kendall, K., Alford, N. McN., Clegg, W. J. and Birchall, J. D., Flocculation clustering and weakness of ceramics, *Nature,* **339** (1989) 130–2.
19. Kendall, K., Macro defect free cements, *Euro. J. Engng Educ.,* **12** (1987) 21–5.
20. Birchall, J. D., Inorganic fibres, *Encyclopedia of materials Science and Engineering*, Pergamon Press, Oxford, 1986, pp. 2333–6.
21. Kendall, K., Alford, N. McN., Clegg, W. J. and Birchall, J. D. Advancing ceramics through agglomerate control, *Institute of Chemical Engineers — 5 Int. Symp. on Agglomeration*, Institute of Chemical Engineers, Rugby, UK, 1989, pp. 355–9.

9

Ceramic Matrix Composites: Properties and Applications

P. DESCAMPS, J. TIRLOCQ AND F. CAMBIER

Centre de Recherches de l'Industrie Belge de la Céramique (CRIBC), 4 Avenue Gouverneur Cornez, 7000 Mons, Belgium

1. INTRODUCTION

Advanced ceramics exhibit a combination of properties: high strength at elevated temperature, high hardness, good corrosion and erosion behaviour, high elastic modulus, low density and generally low coefficients of friction, that make them potential candidates for many structural applications. Today major applications of advanced ceramics include cutting tools, wear components, bioceramics, heat exchangers, coatings, etc. However, to allow their use in new areas such as engines, turbines, etc., it is necessary to improve their reliability and to reduce their brittleness.

One of the most attractive ways to achieve this goal consists of the strengthening and toughening through composite development. Two kinds of ceramic composite are generally considered: dispersion composites and long fibre composites. The former class includes composites with dispersions of:

- particles, chemically and/or crystallographically different from the matrix,
- whiskers, consisting of elongated monocrystals of length in the range from 10 to 100 μm and diameter from 0·1 to 2 μm,
- platelets, of 1 or 2 μm thickness and 10 to a few 100 μm in diameter,
- short fibres (a few mm in length, 10–100 μm in diameter).

Although their toughness is increased, dispersion composites generally show brittle behaviour; conversely, when some requirements are satisfied,[1-3] fibre–matrix composites show non-catastrophic failure.

2. MAJOR FAMILIES OF COMPOSITE

2.1 Long Fibre Composites

Several papers summarize the long fibres commercially available.[4,5] Carbon–carbon fibre composites (C/C) have been intensively studied and have found applications, especially in the aircraft industry as wing, forward fuselage, empennage and brake components. However, C/C composites are strongly sensitive to oxidation and various coatings, impregnants and inhibitors have been tried to avoid carbon gasification. Even though the inhibitor route using organohalogens, phosphates, boron, organoboron compounds or organophosphorus compounds[6] is very promising, at the present time, C/C composites cannot be protected above 1400°C for long periods.

Early work using carbon fibres in ceramic and glass matrices has established the potential offered by fibre reinforcement.[7,8] These composites are however also degraded under oxidizing environments at relatively low temperature and refractory ceramic fibres appear to be most suitable for high temperature uses. SiC fibres have been the most extensively studied and among them, 'Nicalon',[9] produced by Nippon Carbon (Japan), a polycarbosilane-derived microcrystalline silicon carbide (β-SiC) fibre containing oxygen and excess carbon. Nicalon fibre has the special property of favouring the in-situ formation of a carbon rich interface layer. The major consequence of such a matrix–fibre interface is the decrease of interfacial shear stress which gives toughness improvement by fibre debonding and pull-out mechanisms. Compared to SiC fibres, oxide fibres (alumina (Al_2O_3) for example) have received less attention because fibres tend to bond chemically to most ceramic matrices and therefore composites fabricated with such oxide fibres behave like monolithic ceramics. Thus, it is obvious that for fibre composites, the matrix–fibre interface governs the mechanical behaviour.

Difference in thermal expansion coefficient (α) between fibre and matrix may also strongly influence the fibre–matrix interface.[10] For a composite where infiltration of the fibre preform occurs at high temperature, the interface undergoes compressive stresses during cooling when $\alpha_m > \alpha_f$. Consequently fibre debonding and pull-out become non-operative and the composite fails catastrophically. Conversely, when $\alpha_m < \alpha_f$, the interface is submitted to tensile stresses. Thus, the bonding is weakened and the toughening mechanisms become effective.

Glass matrix composites have been preferentially studied because of the possibility of fabricating them by hot-pressing. Indeed, viscous flow of the matrix between fibres aids the easy production of dense materials; such materials have been developed mainly for aircraft applications.[11] Oxide, carbide and nitride matrices have received less attention and whiskers have usually been preferred to continuous fibres. However, alternative fabrication techniques[12-14] including, for instance, organic precursor infiltration, sol infiltration, vapour phase processes (chemical vapour infiltration (CVI) chemical vapour deposition (CVD)) allow extension of the field of fibre/refractory ceramic matrix combinations (SiC–Si_3N_4, Al_2O_3–Al_2O_3, SiC–Al_2O_3, for example).

Ceramic fibre–ceramic matrix composites, due to their reduced brittleness, appear to be suitable materials for structural applications, but their thermal stability is still questionable. For instance, SiC fibres can undergo some oxidation at high temperature. Several papers concern the study of the oxidation behaviour of carbon coated SiC fibre.[4] At high temperature, the carbon layer burns out and is replaced by an amorphous silicate layer resulting from the reaction between the matrix and the oxide fibre, leading to a strong fibre–matrix interface which renders the composite brittle. Consequently, the specific needs to enlarge the applications of continuous fibre–matrix composites require the development of thermally stable fibres compatible with refractory ceramic matrices and the optimization of the ceramic–matrix interface through fibre coating techniques.

2.2 Dispersion Composites

2.2.1 Particle composites

Large improvements in mechanical property have been achieved by dispersion in a ceramic matrix of a second phase that can undergo phase transformation. Zirconia (ZrO_2) based composites belong to this class and have been the most extensively studied. Hafnia (HfO_2) and β-dicalcium silicate dispersions (β-Ca_2SiO_4) have received less attention.

Dispersions of ZrO_2 in alumina and mullite matrices result in large improvements in toughness and, following Irwin's relation $K_{Ic} \sim \sigma_F \cdot a_c^{1/2}$,[15] where a_c is the critical flaw size, in strength (Tables 1 and 2). Possible

TABLE 1
Room temperature mechanical properties of Al_2O_3 matrix composites[21]

Composition	Pressureless sintered		Hot-pressed	
	σ_F(MPa)	K_{Ic}(MPa m$^{1/2}$)	σ_F(MPa)	K_{Ic}(MPa m$^{1/2}$)
A5Z	459	5·85	670	8·1
A10Z	—	5·50	795	10·0
A15Z	380	5·75	750	6·5
A20Z	345	6·05	625	6·0
A20Z1Y	540	5·60	825	8·2
A20Z2Y	557	6·50	1 080	10·1
A20Z3Y	520	5·05	700	8·5
A45Z3Y	655	11·00	1 700	13·5

TABLE 2
Room temperature mechanical properties of mullite/zirconia and mullite/alumina/zirconia composites (prepared by the reaction sintering route)[22]

Composition	Pressureless sintered		Hot-pressed	
	σ_F(MPa)	K_{Ic}(MPa m$^{1/2}$)	σ_F(MPa)	K_{Ic}(MPa m$^{1/2}$)
ZAM 1	270	4·6	—	—
ZAM 2	315	4·75	—	—
ZAM 3	330	5·25	—	—
Mullite	150	1·8	270	—

ZAM 1: 75% mullite–25% ZrO_2.
ZAM 2: 56% mullite–20% Al_2O_3–18% ZrO_2.
ZAM 3: 72% Al_2O_3–11% mullite–17% ZrO_2.

reinforcement mechanisms[16–18] are transformation toughening if the metastable tetragonal phase to monoclinic phase transformation occurs during crack propagation, and microcracking if the phase transformation occurs during cooling. Even if the second mechanism leads to a lesser extent of reinforcement, it is still effective at high temperature where transformation toughening becomes non-operative (the tetragonal phase is stable at high temperature and transformation is inhibited). For Al_2O_3–ZrO_2 composites[10] with low zirconia content (less than 10 vol.% ZrO_2), transformation toughening overcomes microcracking toughening, whereas at higher volume contents (>20 vol. % ZrO_2, the microcracking mechanism becomes dominant.

Mullite-ZrO_2 composites have also been extensively studied;[18-20] however at the present time correlations between parameters associated with the presence of a dispersed phase, and the mechanical properties of the composite, have not been clearly established. The microstructure of the mullite grains may differ depending on the additive[18] (CaO, MgO, Y_2O_3 or TiO_2). Magnesium oxide (MgO) additions, for example, induce the development of a needle-like morphology for the mullite grains, and other reinforcement mechanisms such as crack deflection (see later) will interfere with the mechanism associated with the addition of ZrO_2.

Other composites involving dispersion of 'hard' or 'soft' particles have been developed. Soft particle composites consist generally of metal particle dispersions in a brittle matrix such as, for instance, stainless steel particles in a sialon matrix,[23] for which toughness improvement can reach almost 60% (sialon: $K_{Ic} = 7{\cdot}7$ MPa $m^{1/2}$; sialon/stainless steel particles: $K_{Ic} = 12{\cdot}6$ MPa $m^{1/2}$). In this case, the crack front passes through the stress-free interface and is blunted when it encounters the ductile dispersoid which can deform plastically. Plastic deformation absorbs the energy required for crack propagation and the crack is impeded, resulting in a toughness increase. Metallic dispersoids are, however, because of their poor oxidation resistance, restricted to relatively low temperature of use, so that hard refractory particles are often preferred.

Hard particle composites are numerous and many papers and patents have been published. For instance, in the area of cutting tool applications[24-38] the following systems have been studied: Al_2O_3–SiC, Al_2O_3–TiC, Al_2O_3–TiN, Al_2O_3–Ti(C_x,N_{1-x}), Si_3N_4–TiC, Si_3N_4–sialon. It should be emphasized that wide attention has been given to carbides and nitrides as dispersed phases. In the latter case, these additions improve more or less significantly hardness, fracture toughness, thermal and mechanical shock resistance and thermal and electrical conductivity. This last property is not the least interesting as it allows the possibility of electrical discharge machining (EDM).[39]

Possible toughening mechanisms related to hard particle additions are:

— Microcracking that may exist when the thermal expansion coefficient of the dispersed particles is lower than that of the matrix. Generally the critical particle size for which microcracking occurs is large (several tens of μm) and toughening improvement occurs at the expense of strength. As an example, a Si_3N_4–SiC (32 μm) composite results in an increase of fracture energy,[26] but in a strength loss (by 40%).[40]

— Crack deflection[41,42] that assumes the presence of a stress field surrounding the ceramic–matrix interface and requires thermal expansion or elastic modulus mismatch. Crack deflection produces tilting and twisting of the crack front which increases the crack path length, and thus the fracture energy. In this case, toughening is dependent upon the dispersoid shape. Rod shaped particles (SiC whiskers or short fibres) are predicted to be more effective than disc shaped particles (SiC platelets) which are more effective than spheres (SiC).

— Crack branching and crack bowing[18] for which toughness reinforcement also depends on the increase in crack path length. However, in some particular cases, such as Al_2O_3–TiC composites, none of these mechanisms is effective. Indeed Al_2O_3–TiC composites can have a transgranular fracture mode,[24,25] and therefore the crack passes through the TiC particles and consequently the fracture energy of the composite is a balance of the Al_2O_3 and TiC fracture energies.

2.2.2 *Silicon carbide whiskers (SiC-w)*

Whisker dispersions, mainly in a ceramic matrix (for example Al_2O_3, Si_3N_4, mullite, cordierite) result in an outstanding improvement of the mechanical properties (toughness, strength and creep resistance). This section will be focussed on alumina and silicon nitride-based composites which have been the most widely studied. The largest improvement is obtained for Al_2O_3–SiC-w composites.[43] Addition of 30 vol. % results in a fracture toughness increase at room temperature from 5 MPa $m^{-1/2}$ to 9·5 MPa $m^{1/2}$, which a concomitant increase of bend strength from 385 MPa to 650 MPa. Whisker composites keep their good mechanical behaviour at high temperatures (generally up to 1000 °C).[44] Between 1000 and 1200 °C, the strength drops significantly, whereas a parallel K_{Ic} increase is observed.

Compared to fine-grained alumina, SiC whisker reinforcement was shown to reduce deformation rates during high temperature creep,[45] so that such materials could meet the requirements of many severe applications, with the exception of applications where resistance to thermal shock is necessary (for example engine parts such as combustion liners). For such applications, Si_3N_4–SiC-w composites are considered to be more promising materials. Although improvement is less than for alumina matrix composites, the addition of 30 vol. % SiC-w has been observed to increase toughness by 40% and strength by 25%.[46] Room temperature fracture toughness increases from 4·7 MPa $m^{1/2}$ to 6·4 MPa $m^{1/2}$, with an increase of bend strength from 773 MPa to about 970 MPa. The same material tested at 1000 °C and 1200 °C showed that effective reinforcement is maintained.

From the fracture face analysis[43,46] it appears that both for alumina and for silicon nitride matrix materials crack deflection and whisker pull-out are the main mechanisms which contribute to toughening. Thus, as for toughening in continuous fibres, optimization of reinforcement will require optimization of the fibre–matrix interface by coating of the whiskers. Indeed, C coating of SiC fibres has led to significant improvement of toughness. For alumina based composites,[47] toughness values in the range of 7·5 MPa $m^{1/2}$ were obtained with the as-produced SiC whiskers. By using carbon coated whiskers, the K_{Ic} value increased to 8·7 MPa $m^{1/2}$.

2.2.3 Platelet composites

Platelets appear to be an alternative to reinforcement by whiskers. Platelets have normally a wide size distribution (ranging from 10 to 50 μm) and there is a risk that the coarser platelets will act as critical defects. In this case, toughness improvement should be at the expense of strength. This is why, before mixing, platelets should be separated into narrow sized fractions, to eliminate the coarser particles. Data collected in our laboratory have shown an increase of toughness from 7·05 to 8·5 MPa $m^{1/2}$ and a strength increase from 612 to 671 MPa by incorporating 10 vol. % of narrow size mass SiC platelets ($<20\ \mu$m) in a Si_3N_4 matrix. Moreover, platelets are attractive candidates as ceramic reinforcement materials, because, compared with whiskers, their cost is lower[47] and they are not toxic.[47,48]

3. APPLICATIONS

Ceramic composites are likely to find applications in many fields. Some attractive ones are reviewed in this section.

3.1 Heat Engines

Adiabatic or low heat rejection diesel engines and gas turbines for power generation, and more recently for automotive use, are at the present time receiving considerable attention. Large programes exist in Japan (MITI), USA (DOE) and Germany (the Ministry of Research and Technology) with the aim of developing ceramic gas turbines. Researchers estimate that engines with major ceramic components may improve engine efficiency by 30–50% over those based on current engine technology.[49]

Because of their high toughness, strength and thermal shock resistance, Si_3N_4-based composites such as the Si_3N_4–SiC-w composites are potential candidates for engine applications. In order to fabricate complex shapes, such as turbochargers or gas turbine disks, injection moulding[50] and slip casting techniques[51] have been applied to these composites. However, in both cases, due to preferential orientation of the whiskers, anisotropic shrinkage occurs, and often impedes manufacturing. Moreover, the preferential orientation of whiskers also influences the mechanical behaviour. K_{Ic} anisotropy has been particularly emphasized in hot-pressed Si_3N_4–SiC-w composites.[46]

Due to the high creep resistance of SiC, SiC composites are also planned for the same applications. Thus, IHI (Japan), in the framework of the Japanese gas turbine project, is developing C-coated SiC fibre–SiC matrix composites with the goal of achieving a K_{Ic} of >6 MPa $m^{1/2}$ at room temperature, and a strength of 400 MPa maintained up to 1400°C. The mechanical properties of SiC-based materials[52–54] are also considerably improved by TiB_2 particle additions (10–20 vol. %). Adding TiB_2 to a SiC matrix may increase the mean strength and mean fracture toughness of the composite by 30% and 90% respectively.[54] However, even if the mechanical properties are maintained at 1200°C, a marked decrease occurs at 1400°C, because of catastrophic oxidation.

At the present time, a barrier to the successful wider application of ceramic components is lack of knowledge concerning performance under cyclic stressing.[55] The ceramist's dynamic fatigue test consists essentially of tensile tests performed at various strain rates. These are not sufficient if one takes into account the fact that a high revolution turbine will accumulate 10^9 cycles or more during its lifetime.

3.2 Wear Parts and Cutting Tools

Advanced ceramics and composites are of great interest in industrial applications involving heat, corrosion and wear resistance, and are now used for valve seals, bearings and cutting tools. In the context of the latter application, manufacturing costs make it necessary to find new tools to allow operation at higher cutting speeds. However, these conditions induce higher stresses and increase the temperature at the tool–workplace interface. Monolithic ceramic materials (mainly alumina), because of their higher wear resistance than cemented carbides or TiC type cermets, have replaced conventional tools for specific machining operations, but their low toughness limits their use. Thus they have been successfully applied to cast iron machining but not for steel; moreover, operations involving interrupted cuts are ruled out. Solutions to these problems have been found through the composite approach.

Al_2O_3–TiC,[30] Al_2O_3–TiN,[36–39] and Al_2O_3–Ti(C_x–N_{1-x})[41] composites have been patented in cast iron machining and their use has been extended to steel and ferrous workpieces. Because of the considerable toughness improvement, SiC-w reinforced alumina has been recently added to the cutting tool arsenal and it is considered to be the state-of-the-art material for machining nickel-based superalloys.[56] In order to avoid the loss of efficiency of Al_2O_3–ZrO_2 cutting tools with increasing temperature, which limits the speed/feed/depth of cut, SiC whiskers have been incorporated into Al_2O_3–ZrO_2 materials.[57]

Next to the alumina-based composites, Si_3N_4 composites and especially Si_3N_4–TiC, Si_3N_4–sialon and Si_3N_4–SiC-w composites are also of considerable interest for cutting applications.[58] Coatings have also been investigated as a way of improving the tribological performance of tools. For instance, high cutting speeds cause welded adhesion between ferrous workpieces and SiC-w reinforced tools during machining. This kind of problem can be overcome by coating the tool with thin films of TiC and TiN.[59]

3.3 High Temperature Corrosion

Good thermal shock resistance and corrosion resistance to molten metals are necessary for steel making, especially in the horizontal continuous casting process. Ceramic composites containing boron nitride (BN)[10] (for example Al_2O_3–BN, Si_3N_4–BN) satisfy these two conditions and may be considered as potential candidates for replacing expensive bulk BN materials. In another application, heat exchangers that use hot exhaust gas to preheat inlet air or gas, require materials having good thermal conductivity, good thermal shock resistance and good resistance to corrosive environments. The ceramic currently used in this field is silicon carbide.[60] However, aluminium nitride and silicon nitride, and composites such as Si/SiC, may be expected to find applications in this field in the future.[49]

3.4 Medical Products

Based on their inertness and wear resistance compared to titanium and alloys, alumina and alumina-based composites (Al_2O_3–ZrO_2) have been used *in vitro* and *in vivo* in research for applications including orthopaedic knees, hip joint prostheses, tooth implants and bone implants.[49]

4. PROCESSING OF COMPOSITES

The processing techniques for dispersion composites are similar to those involved in the fabrication of bulk ceramics. However, the presence of the dispersed phase inhibits or reduces sinterability. To obtain fully dense materials it is often necessary to increase the diffusion rates of species by increasing the volume of liquid phase, or by applying external forces such as those of hot-pressing or hipping. It is not the aim of this paper to describe well-known ceramic procedures. In the following sections, new techniques that can be applied to composites with the goal of improving properties and/or decreasing production costs, are briefly introduced.

4.1 Solution Chemistry Processing[20, 61] (sol–gel, solution-precipitation)

This provides the opportunity for developing new powders (of controlled particle size, shape, purity and uniformity) that enables the mechanical characteristics of the final ceramic components to be improved, especially at high temperature. Thus, pure mullite–zirconia composites made by the sol–gel route maintain their room temperature strength and fracture toughness to 1200 °C.

4.2 Microwave Processing[62] and Self Propagation High Temperature Synthesis (SHS[63])

These appear particularly attractive alternative sintering routes, in order to reduce markedly sintering time and cost. In the SHS process, the heat generated in exothermic reactions is used to decrease the required production energy.

4.3 Tape Casting[64]

This has also attracted considerable interest. A very attractive feature of this type of structure is that several property requirements can be satisfied simultaneously (that is, corrosion resistant surfaces, tough matrix) and the final composite tape obtained by this method may be very thin (<1 mm). The technique generally involves producing laminates from polymer-based suspensions of ceramic, stacking the layers in a prescribed sequence to form a green laminate, and sintering the laminate in successive stages to produce a ceramic-based laminated composite.

4.4 Superplastic Forming

Following the discovery of the superplasticity of ceramics, superplastic joining by solid state bonding also allows[65] multilayer composite fabrication to be envisaged. In this case a strong bonded interface may be obtained if at least one of the materials to be joined is superplastic, and if their thermal expansion coefficients are close. Another interesting feature of superplasticity is that it permits the extension of hot-forging techniques to ceramics (for example, superplastic forging and bulge forging). The feasibility of these techniques has already been demonstrated on the laboratory scale for tetragonal zirconia polycrystal (TZP) and ZrO_2 particulate composites.[66]

4.5 Surface Treatment

In addition to these 'bulk' processing techniques, the surface treatment of ceramics (by ion implantation or laser methods) is also of great interest.[49] For instance, laser treatment of ceramics may promote or improve densification and increase hardness and toughness (for example ZrO_2 combustion liner coatings prepared by plasma spraying).

4.6 Long Fibre Composite Processing

For continuous fibre composites, the fibre structure impedes shrinkage, and other techniques must be involved. Woven fibre preforms are made by methods currently used in the textile industry,[67] and the matrix is introduced by infiltration or deposition techniques. The main techniques used are reported here.

4.6.1 Infiltration techniques

Slurry infiltration.[11] Fibres are impregnated by a slurry and then hot-pressed. For complete densification a temperature just above the softening point of the matrix is required. This technique is in current use for glass ceramic fabrication. In the case of crystalline matrices, the method appears less attractive because of the high densification temperature which can damage fibres or induce chemical reactions at the fibre–matrix interface.

Organic precursor infiltration.[68] The processing consists of the infiltration of a preform by organic precursors (polymers, or pitch, in the case of C-C composites) which after heat treatment will form the matrix. The technique allows low temperature processing. However, the important shrinkage during sintering requires care to be taken in order to avoid cracking of the matrix, and impregnation of the fibres takes place in several steps.

Sol-gel process.[69] The fibre preform is infiltrated by a sol. After drying, the gel forms and is heat treated to produce the matrix. The advantages and problems of the technique are similar to those of the organic precursor infiltration method.

Melt infiltration technique.[64] This consists of the infiltration of a fibre preform by a molten matrix, followed by hot-pressing. This method is close to the squeeze casting technique used for organic and metallic matrices, and can only be applied if the melting point of the matrix is sufficiently low. The main advantage of the technique is that, due to small dimensional changes, it may be used to manufacture final shaped components. If the infiltration temperature is too high, fibre damage may occur. Moreover, this general method generally gives a strong bonding between the fibre and the matrix.

Liquid reaction with a preform.[5] The preform is impregnated by a liquid which can then react with the fibre to form the ceramic material. This process leads to dense materials but often requires high temperature, and fibres are modified as a consequence of the reaction.

Deposition techniques.[70] These consist of chemical vapour infiltration or deposition on a fibre preform (CVI or CVD). A flow of reactive gas passes through a heated fibrous preform and condenses around the fibre to form the matrix. Because of the low deposition rates on the matrix (a few μm h^{-1}), the process is expensive and can not be applied to large scale markets.

5. CONCLUSION

Due to their improved mechanical properties compared with monolithic ceramics, ceramic composites are promising materials, mainly for high temperature structural applications. They have received a great deal of interest, especially for wear resistant components, cutting tools and, more recently, as engine components.

Extensive development of these ceramics depends strongly on the three following conditions:

— The reduction of the manufacturing costs for powders and final products. Some techniques briefly described in this paper (for example SHS and microwave sintering) appear to be attractive as routes to achieving this goal.

— The improvement of the reliability of ceramic parts in use, which requires optimization of toughness. This could be successfully achieved through a better understanding of the microstructure/mechanical properties relationship, for example by determination of the operating toughening mechanisms and by control of the fibre–matrix interface in the case of fibre and whisker composites. In the same way, a major effort must be made to develop suitable cyclic fatigue tests.

— The control of the transfer of technology from laboratory to industrial-scale production. It is well established that more than 10 years are necessary to take materials from the laboratory to the industrial stage. It is also known that such transfer is easier in Japan than in the USA or Europe, which could be one of the reasons for the leading position of Japan in the advanced ceramic field.

REFERENCES

1. Rouby, D. and Navarre, G., Interfaces et Micromechanisms dans les Composites Fibreux à Matrice Ceramique, *Silicates Industriels* **55** (7/8) (1990) 201–16.

2. Sheppard, L. M., A global perspective of advanced ceramics, *Am. Ceram. Soc. Bull.*, **68** (9) (1989) 1624–33.
3. Strong growth predicted for several ceramic markets, *Am. Ceram. Soc. Bull.*, **67** (12) (1989) 1888–9.
4. Mah, T., Mendirata, M. G., Katz, A. P. and Mazdiyasni, K. S., Recent developments in fibre-reinforced high temperature ceramic composites, *Am. Ceram. Soc. Bull.*, **66** (2) (1987) 304–8.
5. Rouby, D., Les matériaux composites à fibres et matrice céramique. *Proceedings of the IIme Conference Franco-Allemande sur les céramiques techniques*, Aachen, 4–6 March 1987. Institut für Gesteinshüttenkunde der RWTH-Aachen, 1987, pp. 265–86.
6. Sheppard, L. M., Challenges facing the carbon industry, *Am. Ceram. Soc. Bull.*, **67** (12) (1988) 1897–902.
7. Sambell, R. A. J. *et al.*, Carbon fibre composites with ceramic and glass matrices, Part 2: Continuous fibres, *J. Mater. Sci.*, **7** (6) (1972) 676–81.
8. Phillips, D. C., Sambell, R. A. J. and Brown, D. H., The mechanical properties of carbon fiber reinforced pyrex, *J. Mater. Sci.*, **7** (1972) 1454–64.
9. Yajima, S., Okamura, K., Hayashi, J. and Omori, M., Synthesis of continuous SiC fibers with high tensile strength, *J. Am. Ceram. Soc.*, **59** (7–8) (1976) 324–7.
10. Fantozzi, G., Comportement mécanique des céramiques composites à fibres et à dispersoides, *Silicates Industriels*, **53** (5–6) (1988) 67–84.
11. Moore, R. E., Long, M. C. and Ha, J., Development of glass ceramic matrix composites for aircraft application. Presented at the 2nd Int. Conf. on Ceramic–Ceramic Composites, Mons, Belgium, 17–19 October 1989. To be published in *Silicates Industriels*, **56** (1991).
12. Cornie, J., Chiang, Y. M., Uhlmann, D. R. and Mortensen, A., Processing of metal and ceramic matrix composites, *Am. Ceram. Soc. Bull.*, **65** (2) (1986) 293–304.
13. Schioler, L. J. and Stighlich, J. J., Ceramic matrix composites — a literature review, *Am. Ceram. Soc. Bull.*, **65** (2) (1986) 289–92.
14. Cales, B., Ceramic matrix composites. In: *Proceedings of the 2nd European Symposium on Engineering Ceramics*, Riley, F. L. (ed.), Elsevier Applied Science, London, 1989, pp. 171–202.
15. Jayatilaka, A., *Fracture of engineering brittle material*. Elsevier Applied Science, London, 1989, p. 21.
16. Evans, A. G., Toughening mechanisms in zirconia alloys, *Adv. Ceram.*, **12** (1984) 193.
17. Lange, F. E., Transformation toughening — Parts I to IV, *J. Mater. Sci.*, **17** (1982) 225–54.
18. Leriche, A., Influence des paramètres d'élaboration de composites mullite–zircone sur leur microstructure. PhD thesis, Université de l'Etat, Mons, Belgium, 1986.
19. Leriche, A., Descamps, P. and Cambier, F., High temperature mechanical behaviour of mullite zirconia composites obtained by reaction sintering. *Zirconia '88, Advances in Zirconia Science and Technology*, eds. B. Meriani and C. Palmonari. Elsevier Applied Science, London 1989, pp. 137–51.

20. Grahl-Madsen, L., Daugaard, C., Engell, J., Leriche, A., Descamps, P. and Cambier, F., Zirconia toughened mullite ceramics prepared from alkoxides. *Silicates Industriels* **55** (9/10) 247–57.
21. Leriche, A., Moortgat, G., Cambier, F., Homerin, P., Thevenot, F., Orange, G. and Fantozzi, G., Preparation and microstructure of zirconia-toughened alumina ceramics, *Adv. Ceram.*, **24** (1988) 1033–41.
22. Orange, G., Fantozzi, G., Cambier, F., Leblud, C., Anseau, M. R. and Leriche, A., High temperature mechanical properties of reaction-sintered mullite–zirconia and mullite/alumina/zirconia composites, *J. Mater. Sci.*, **20** (1985) 2533–40.
23. Edrees, H. J. and Hendry, A., Metal reinforced ceramic matrix composites. *Silicates Industriels* **55** (7/8) (1990) 217–22.
24. Wahi, R. P. and Ilschner, B., Fracture behaviour of composites based on Al_2O_3–TiC, *J. Mater. Sci.*, **15** (1980) 875–85.
25. Wahi, R. P., Fracture behaviour of two-phase ceramic alloys based on aluminium oxide, *Trans. Indian Inst. Metals*, **34** (2) (1981) 89–102.
26. Grellner, W., Hubner, H., Ilschner, B. and Kleinlein, F. W., On high temperature strength of a two-phase Al_2O_3 base material, *Sci. Ceram.*, **10** (1980) 513–19.
27. Lee, S. H., Ceramic compositions. Int. Patent No. WO 81/01143, 1981.
28. Lee, S. H., Wear resistant ceramic materials. Int. Patent No. WO/01144, 1981.
29. Borom, M. P. and Lee, M., Effect of heating rate on densification of alumina–titanium carbide composites, *Adv. Ceram. Mater.*, **1** (4) (1986) 335–40.
30. North, B., Ceramic cutting tools — a review, *Int. J. High Tech. Ceram.*, **3** (1987) 113–27.
31. Yoshimura, H., Ito, N., Nishigaki, K. and Anzai, K., Metallkeramik für Schneidwerkzeuge und daraus hergestellte Schneidplattchen. Patent (D) No. DE 3346873A1, 1984.
32. Tanaka, H., Yamamoto, Y. and Sakurai, K., Keramischer Formkörper für die spanende Bearbeitung und verfahren zu seiner Heistellung. Patent (D) No. DE 2919370C2, 1983.
33. Tanaka, H. and Yamamoto, Y., Gesinterde Keramik insbesondere für Zuspannungswerkzeuge und Verfahren zur Herstellung derselben. Patent (D) No. DE 3010545A1, 1980.
34. Tanaka, H. and Yamamoto, Y., Gesinterter Keramikkörper für Schneidwerkzeuge und Verfahren zu dessen Herstellung. Patent (D) No. DE 3027401A1, 1981.
35. Katsumura, Y. and Fukuhara, M., Plastic deformation in Al_2O_3–Ti(C_x, N_{1-x}) ceramics, *High tech. ceramics*, Vincenzini, P. (ed.), Elsevier Science Publishers, London, 1987.
36. Aspinwall, D. K., Tunstall, M. and Hummerton, R., Cutting tool life comparisons. *Proceedings of the 25th Int. Machine Tool Design and Research Conference, Tobias, S. A. (ed.), University of Birmingham, 1985, pp. 269–77.*
37. *Laugier, M. T., Surface toughness of ceramics, J. Mater. Sci.*, **5** (1986) 252.
38. Furakawa, M., Nakamo, O. and Takashima, Y., Fracture toughness in the system Al_2O_3–TiC ceramics, *Nippon Tungsten Rev.*, **18** (1985) 16–22.

39. Kamiyo, E., Honda, M., Takeuchi, H., Higuchi, M. and Tanimura, T., Electrical discharge machinable Si_3N_4 ceramics, *Sumitomo Electric Technical Review*, **24** (1985) 183–90.
40. Lange, F. F., Effect of microstructure on strength of Si_3N_4–SiC composite system, *J. Am. Ceram. Soc.*, **56** (9) (1973) 445–50.
41. Faber, K. T. and Evans, A. G., Crack reflection process. I. Thory, *Acta Metall.*, **31** (4) (1983) 565–76.
42. Faber, K. T. and Evans, A. G., *Acta Metall.*, **32** (4) (1983) 577–84.
43. Homeny, J., Waughn, W. L. and Ferber, M. K., Processing and mechanical properties of SiC-Whisker–Al_2O_3-matrix composites, *Am. Ceram. Soc. Bull.*, **66** (2) (1987) 333–8.
44. Wei, C. C. and Becher, P. F., Development of SiC-whisker reinforced ceramics, *Am. Ceram. Soc. Bull.*, **64** (2) (1985) 298–304.
45. Chokshi, A. H. and Porter, J. R., Creep deformation of an alumina matrix composite reinforced with silicon carbide whiskers, *J. Am. Ceram. Soc.*, **68** (6) (1985) C144–5.
46. Buljan, S. T., Baldoni, J. G. and Huckabee, M. L., Si_3N_4–SiC composites, *Am. Ceram. Soc. Bull.*, **66** (2) (1987) 347–52.
47. Black, J. A., Shaping reinforcement for composites, advanced materials and processes, *Metal Progress*, **3** (1988) 51–4.
48. Birchall, J. D., Stanley, D. R., Mockford, M. J., Pigott, F. and Pinto, P. J., Toxicity of silicon carbide whiskers, *J. Mater. Sci. Lett.*, **7** (1988) 350–2.
49. Mecholsky, J. J. Jr, Engineering research needs of advanced ceramics and ceramic-matrix, *Am. Ceram. Soc. Bull.*, **68** (2) (1989) 367–75.
50. Kandori, T., Ukyo, Y. and Wada, S., Directly HIP SiC whisker reinforced Si_3N_4 in whisker and fiber toughened ceramics. In *Proceedings of an International Conference, Oak Ridge, Tennessee, USA, 7–9 June*, eds. R. A. Bradley, D. E. Clark, D. C. Larsen and J. O. Stiegler. ASM International, USA, 1988, pp. 125–9.
51. Hoffman, M. J., Nagel, A., Greil, P. and Petzow, G., Slip casting of SiC-whisker reinforced Si_3N_4, *J. Am. Ceram. Soc.*, **72** (5) (1989) 765–9.
52. Janney, M. A., Mechanical properties and oxidation behaviour of a hot pressed SiC-15 vol% TiB_2 composite, *J. Mater. Sci.*, **25** (1990) 157–60.
53. Tani, T. and Wada, S., SiC matrix composites reinforced with internally-synthesized TiB_2. 13th Annual Conference on Composites and Advanced Ceramic Materials, Cocoa Beach, FL, USA, January 1989; to be published in the Proceedings.
54. McMurty, C. H., Boecker, W. D. G., Seshadri, S. G., Zanghi, J. and Garnier, J. E., Microstructure and material properties of SiC–TiB_2 particulate composites, *Am. Ceram. Soc. Bull.*, **66** (2) (1987) 325–9.
55. Ford, R. G., A development engineer's view of barrier to success in marketing structural ceramics for engine applications, *J. Aust. Ceram. Soc.*, **23** (1) (1987) 47–8.
56. Billman, E. R., Mehrotra, P. K., Shuster, A. F. and Beeghly, C. W., Machining with Al_2O_3–SiC whisker cutting tool, *Ceram. Engng Sci. Proc.*, **9** (7–8) (1988) 543–52.
57. Claussen, N. and Petzow, G., Whisker reinforced oxide ceramics, *J. Phys. C1* (1986) 693–702.

58. Buljan, S. T., Pasto, A. and Kim, H. J., Ceramic whisker and particulate composites: properties, reliability and applications, *Am. Ceram. Soc. Bull.*, **68** (2) (1989) 387–94.
59. Suzuki, J. and Sakakibara, S., Material for cutting tools, use of saw and cutting tools. European Patent No. EP 0247630A2, 1987.
60. Das, S. and Randall, T., Ceramic heat exchangers: cost estimates using a process-cost approach, *Am. Ceram. Soc. Bull.*, **67** (10) (1988) 1684–9.
61. Ramme, R. and Hausner, H., Mechanical properties of ZrO_2 (2% Y_2O_3) derived from freeze dried coprecipitated hydroxides, *Ber. D.K.G.*, **64** (1–2) (1987) 12–14.
62. Sutton, W. H., Microwave processing of ceramic materials, *Am. Ceram. Soc. Bull.*, **68** (2) (1989) 376–86.
63. McCauley, J. W., Some considerations for the evolution of advanced ceramics, James I. Mueller Memorial Lecture, *Ceram. Engng Sci. Proc.*, **9** (7–8) (1988) 553–60.
64. Adair, J. H., Anderson, D. A., Dayton, G. O. and Shrout, T. R., A review of the processing of electric ceramics with an emphasis on multilayer capacitor fabrication, *J. Mater. Edu*, **9** (1–2) (1987) 71–118.
65. Nagono, T., Kato, H. and Wakai, F., Solid state bonding of superplastic material. *Proceedings of the MRS Int. Meeting on Adv. Mat.*, Vol. 7, *Superplasticity*, M. Doyamo, S. Somiya and R. P. H. Chang (eds), Materials Research Society, 1989, pp. 285–92.
66. Wakai, F., Superplasticity of Zirconia toughened ceramics, PhD thesis, Kyoto University, Japan, 1988.
67. Ko, F. K., Preform fiber architecture for ceramic-matrix composites, *Am. Ceram. Soc. Bull.*, **68** (2) (1989) 401–14.
68. Rand, B., Fabrication of carbon–carbon composites by liquid infiltration. Presented at the 2nd Conference on Ceramic–Ceramic Composites, Mons, Belgium, 17–19 October 1989. To be published in *Silicates Industriels*, **56** 1991.
69. Pierre, A. C., Uhlmann, D. R. and Hordonneau, A., Ceramic composites made by sol-gel processing. *Rev. Int. Hautes Temperatures Refract.*, **23** 1986 29–35.
70. Cornie, J. A., Chiang, Y. M., Uhlman, D. R., Mortensen, A. and Collins, J. M., Processing of metal and ceramic matrix, *Am. Ceram. Soc. Bull.*, **65** (2) (1986) 293–303.

10

Ceramic Heat Exchangers for Domestic and Industrial Applications

J. HEINRICH, J. HUBER, H. SCHELTER

Hoechst CeramTec AG, Werk Selb, Wilhelmstr. 14, 8672 Selb, Federal Republic of Germany

R. GANZ AND O. HEINZ

Hoechst AG, FTT Neue Technologien, Postfach 80 03 20, 6230 Frankfurt/Main, Federal Republic of Germany

1. INTRODUCTION

There are two ways in which ceramic materials may be of interest for the design of heat exchangers. In the high-temperature range above 800°C new opportunities open up for the recovery of heat energy from waste heat in industrial processes. On the other hand the high corrosion resistance of ceramic materials offers alternative possibilities in the low-temperature range as condensing heat exchangers. Increasing the service life and carrying out processes which are not possible or not economic with conventional materials are further arguments in favour of the application of ceramics in heat exchanger design.[1] Opportunities lie, for example, in widening the temperature limits of metals, graphite, of PTFE and glass.

Because of its 100% leaktightness and its corrosion resistance, silicon infiltrated silicon carbide (SiSiC) is of particular importance. In contrast to metals no significant reduction in strength occurs for SiSiC up to approx. 1400°C (Fig. 1). On the other hand ceramic materials are by nature brittle. At the design stage it is necessary, for example, to prevent tensile stresses, avoid point applications for loads and where possible convert forces into compressive stresses. The ceramic production technique permits only relatively limited component sizes. In the

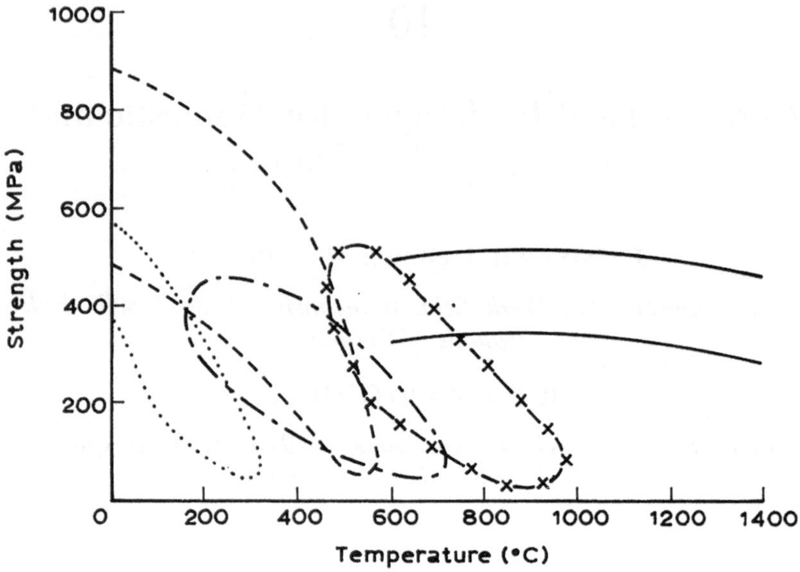

Fig. 1. Strength of (——) silicon infiltrated silicon carbide compared with (- - -) titanium, (· · ·) aluminium, (- · - ·) stainless steel and (-X-X) superalloys.

production of functional units for industrial applications, complex joining and connecting techniques are therefore necessary. The ceramic material silicon infiltrated silicon carbide (SiSiC) permits in combination with the tape casting technique the design of compact heat exchangers with highly complex structures. In many cases, flue gas streams contain particles, for example in forging furnaces. A cleaning facility must be available for dust-laden flue gas streams. With liquid/liquid and liquid/gas heat exchange it may be meaningful to alter the transfer surface, install turbulence generators, etc., and hereby establish process optimizing designs. The tape casting technique described here offers the possibility for doing so and provides the design with greater opportunities than with any other ceramics production technique.

This paper will describe the efficiency, application and production of ceramic compact heat exchangers made from SiSiC. This is first a plate-type cross-flow design with external dimensions of 300 × 300 × 150 mm^3 and rectangular channels. The appropriate function for this type of element is the preheating of combustion air from corrosive flue gases with temperatures above 800 °C. When metal heat exchangers are used, the hot exhaust gases are generally quenched by air to the permissible

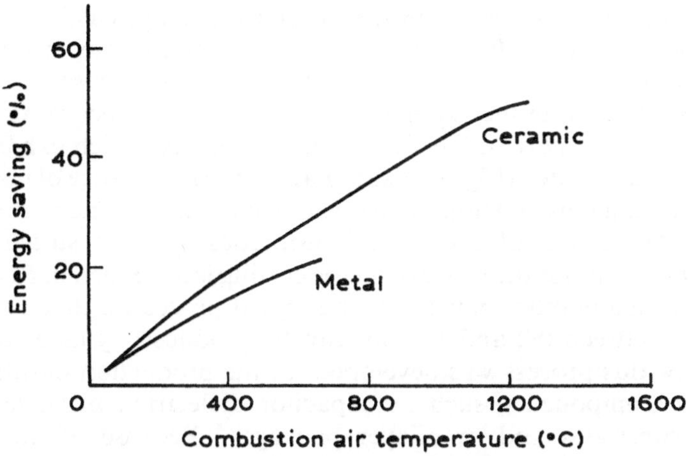

Fig. 2. Energy savings by preheating combustion air.

heat exchanger inlet temperature of approx. 800 °C, thus destroying a high energy potential. The utilization of exhaust gas energy between 800 and 1350 °C permits additional fuel savings of the order of 15–20% (Fig. 2).

The second design described in this paper can be used for a condensing heat exchanger. In order to obtain specific information on the efficiency of these ceramic compact heat exchangers, they must be tested under service conditions.

In a specially equipped test facility the heat transfer characteristics of the cross-flow heat exchanger elements have been determined as a function of various parameters. Results with gas/gas heat exchangers are shown and illustrative examples are given for the use of ceramic heat exchangers in different fields of industrial applications.

In another example the applications of a SiSiC counter-flow heat exchanger in a domestic condensing boiler is described. Information data for a 20 kW unit are presented.

2. PROCESSING TECHNIQUE FOR CERAMIC HEAT EXCHANGER ELEMENTS

For the production of ceramic heat exchangers with a high specific heat transfer surface, tape casting is a convenient moulding process. This technique permits the production of very thin, flat, large-area tapes and based on this, complex structures. Essentially, the tape casting process consists of suspending fine powders in organic or aqueous media with the use of binders and plasticizers, and the casting of these slips on a moving surface. After the vaporization of the solvents according to the binder systems a more or less flexible tape remains, which can be cut, punched or laminated (Fig. 3). A summary of the huge range of possible systems for aqueous and non-aqueous solvents can be found in Ref. 2. A detailed description of rheological properties and the structure of organic materials for the process of tape casting is to be seen in Ref. 3. In contrast to many other powder metallurgical processes, sheets with a thickness between 0·2 and 1·5 mm can be produced by tape casting. Originally, this process was developed for the production of different electronic components, such as capacitor dielectrics, piezoelectrics, ferrites, substrates and multilayer packages.[4] Because of the large surfaces obtainable by means of laminating or winding embossed or punched tapes, substrates for catalysts[5] and heat exchangers can also be produced.

For heat exchangers produced from silicon infiltrated silicon carbide, the starting material is silicon carbide (SiC) powder. After the moulding and the burning out of the organic components the

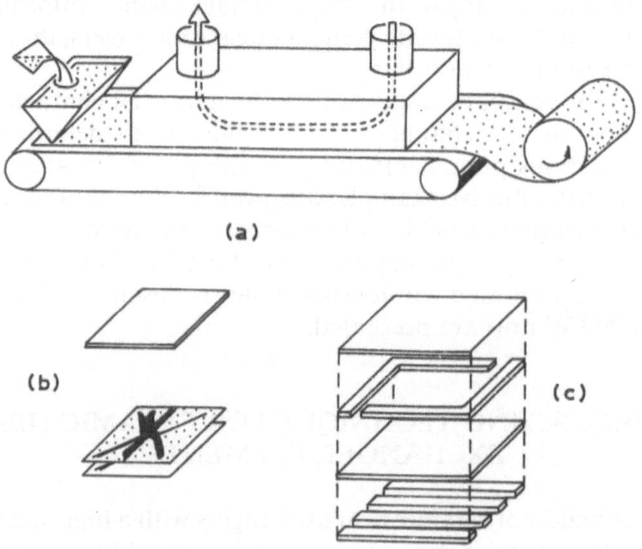

Fig. 3. The tape casting technology. (a) Casting, (b) screen printing, (c) laminating.

remaining pores are filled with silicon. This can either be effected through the gas phase or through the liquid phase. During this process no appreciable change in the overall dimensions can be observed. SiSiC is a dense material consisting of about 90% silicon carbide and 10% silicon. A detailed description of the different processing steps is given in Ref. 6.

3. DESIGN OF HEAT EXCHANGER ELEMENTS

Figure 4 shows the construction of a typical heat exchanger element with rectangular channels in a cross-flow system. The heating unit is fabricated by laminating together two rippled basis plates. The media flow cross-wise through the block element.

For chemical appliances it is sometimes better to have a vairable proportion of the areas of both sides of a heat exchanger. On the other hand round channels can be cleaned easier than rectangular channels. The advantage of a round tube is also a higher bursting pressure. Those features can be achieved, if the heat exchanger element is not fabricated

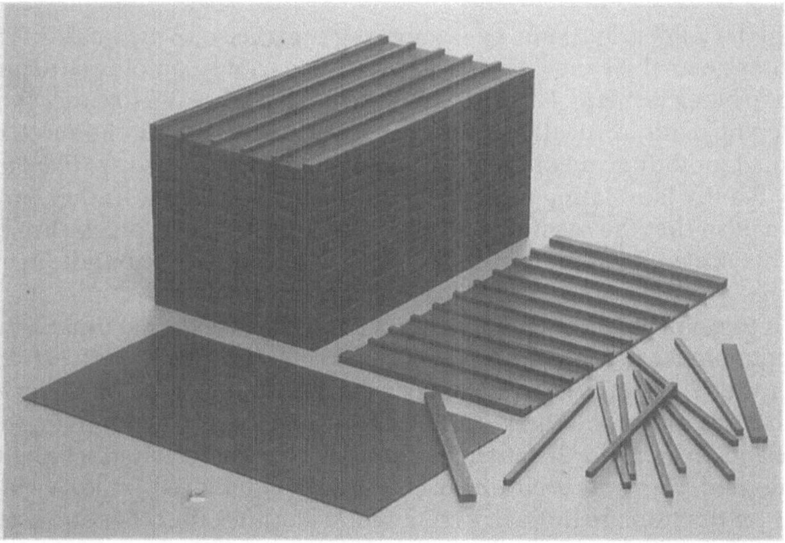

Fig. 4. Construction of a ceramic cross flow heat exchanger with rectangular channels by laminating tapes and stripes.

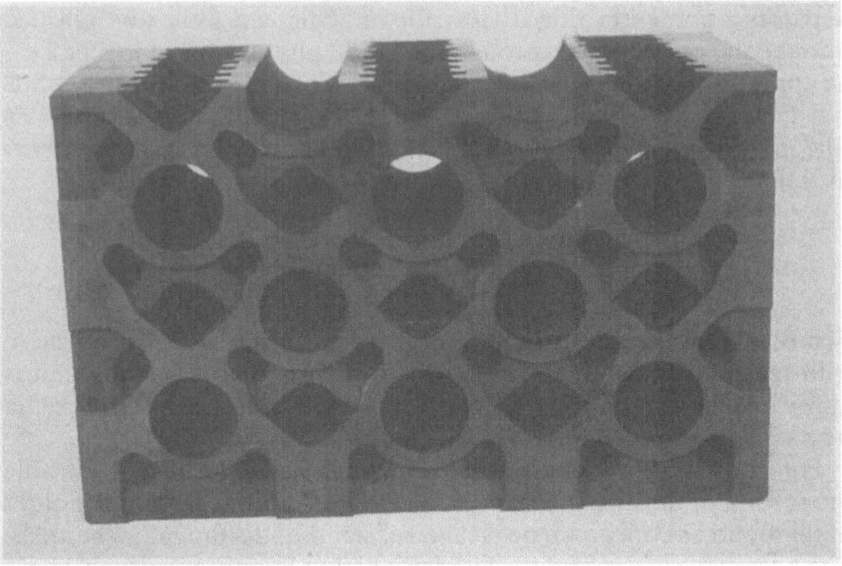

Fig. 5. Construction of a ceramic heat exchanger by laminating punched cards.

by plates and stripes, but by laminating together punched cards. This technique enables the designer to produce various complex structures in one block element. Figure 5 shows that by laminating together those different cards vertically to the card area, tube shaped channels are formed for the one medium. Additionally in this card gaps are fixed in a way that by laminating together the cards in two levels a further gap is formed so that the second medium flows around those 'pipes' through the block element. This can be described as a rippled pipe bundle in one block.[7]

In Fig. 6 the two different types are shown. The external dimensions of both blocks are the same, i.e. $300 \times 300 \times 150$ mm^3. The left element with the rectangular channels has a heat exchange area of 1·5 m^2 on both sides, while the right element has an area of 0·52 m^2 on the tube side and an area of 2·27 m^2 on the slit side. When both constructions are combined, elements with linear channels are achieved on both sides. One product stream flows in a rectangular channel, the other one in any other channel shape.

Figure 7 shows as example a rippled basis plate for a gas/liquid element. For fabricating the block element the basis plates are

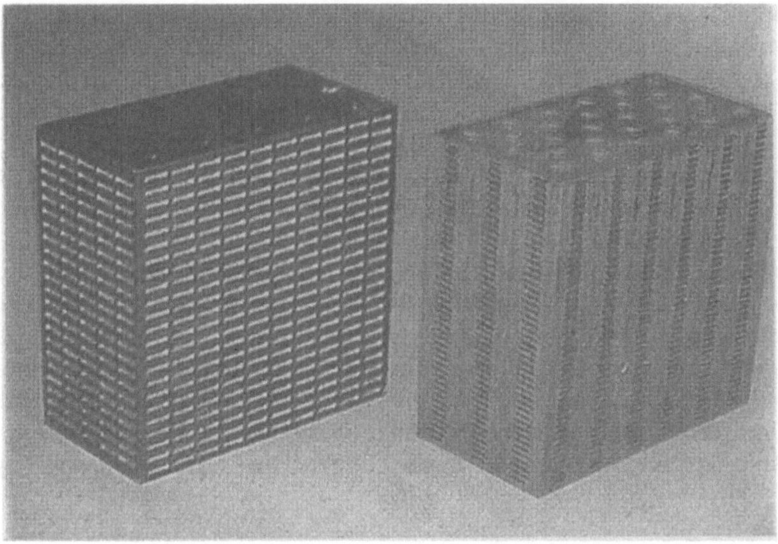

Fig. 6. Different heat exchanger designs.

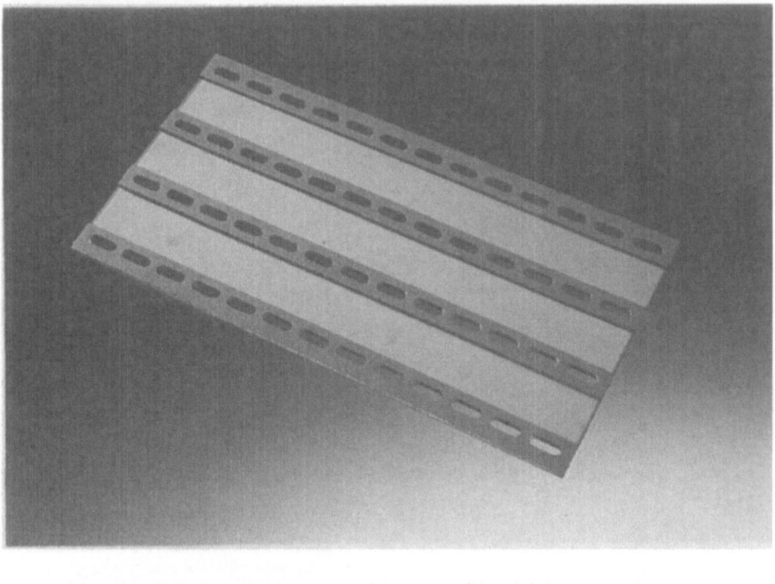

Fig. 7. Rippled basis plate for a gas/liquid heat exchanger.

laminated together in the same direction. The liquid flows through the channels and reaches the gas in the cross-stream. In this case an open cross-stream block element with approx. 4 m^2 gas and 0·6 m^2 liquid contact areas result. The connections for the liquid and the distribution pipes can also be integrated into the ceramic block element.

Figure 8 shows the same element type in three different sizes. The connections for the liquid as well as the distribution pipes are integrated into the element.

For domestic application the boiler part is a ceramic gas–liquid counterflow heat exchanger (Fig. 9). It is also made of SiSiC and fabricated by an assembly of ceramic tapes of 0·8 mm thickness and different strips between the tapes. The flue gas channels and the water channels are straight in the major part of the heat exchanger. Near the water outlet, for example, the entrance of the hot combustion gas, an intensive heat transfer from the walls to the heating water is achieved by

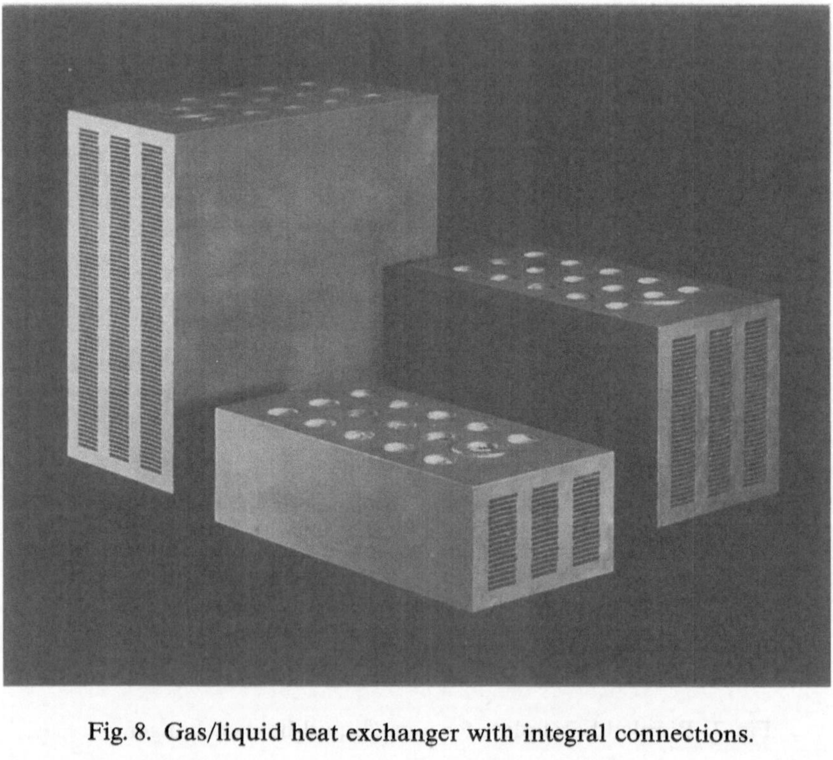

Fig. 8. Gas/liquid heat exchanger with integral connections.

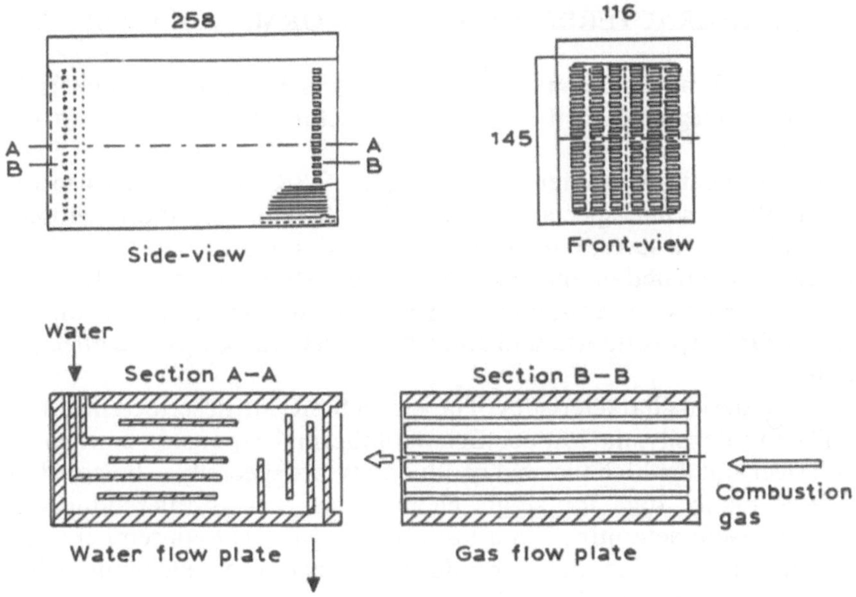

Fig. 9. Basis design of a gas/water counter flow heat exchanger.

the cross-flow pattern. Thus the maximum material temperature will not exceed about 150 °C. Some typical heat exchanger data are shown in Table 1.

The layout of this heat exchanger has been developed in cooperation with the KFA Jülich.[8]

TABLE 1
Characteristic data for a counter-flow heat exchanger for a condensing boiler

Characteristics	Unit	Value
Total volume	dm^3	4·85
Water content	dm^3	0·87
Geometric heat transfer surface	m^2	0·9
Hydraulic diameter of flow channels		
Water	mm	3·15
Combustion gas	mm	4·0
Total weight	kg	7·6
Height, width, depth	mm	258/145/116

4. CHARACTERISTICS AND PERFORMANCE DATA

Silicon infiltrated silicon carbide (SiSiC) is a composite material. Silicon is integrated in the SiC microstructure; the material has no porosity.

For application in chemical apparatus SiSiC was tested as to its suitability in some special media (Table 2). The conditions chosen normally lead to severe corrosion when metals are used. The yearly loss in mm is calculated on the basis of two-week short-time tests. Here the excellent corrosion resistance of SiSiC against aggressive media is shown. Only hydrofluoric acid and 50% $NaOH^-$ cause any real effect to SiSiC.

The material characteristics of tape-cast SiSiC are described in Table 3. The low density, the low coefficient of thermal expansion compared with steel and the high thermal conductivity, are specially to be noted.

Performance data for single elements as well as interconnected elements were determined in a test rig.[9] For gas/gas high-temperature use, the results for the ceramic heat exchanger elements with rectangular channels (4·8 × 23 mm) are indicated. Geometric data for this type of element are shown in Table 4.

TABLE 2
Corrosion of SiSiC by different media

Medium	Temperature (°C)	Weight Loss (mm year^{-1})
Suphuric acid	200	0·018
Hydrochloric acid	180	0·019
Nitric acid	200	0·007
Acetic acid + 3% anhydride	Boiling point	0·003
Acetic acid + 3% anhydride + 100 ppm chloride as NaCl	Boiling point	0·005
Formic acid	Boiling point	0·005
Formic acid + 100 ppm chloride as NaCl	Boiling point	0·004
Hydrofluoric acid	Boiling point	5·49
Sodium hydroxide 50%	150	Dissolved
Deionized water	200	0·304
Deionized water + 100 ppm chloride as NaCl	200	0·288

TABLE 3
Properties of tape-cast SiSiC

Property	Unit	Value
Density	$Mg\ m^{-3}$	3·0
Porosity	%	0
Gas permeability (tape)	$kg\ cm^{-3}\ s^{-1}$	0
Thermal conductivity at RT	$W\ m^{-1}\ K^{-1}$	120
Max. service temp.		
(oxidizing)	°C	1 400
(reducing)	°C	1 400
Specific thermal capacity (20..1000°C)	$J\ kg^{-1}\ K^{-1}$	950
Coefficient of linear expansion (20..1000°C)	$10^{-6}\ K^{-1}$	4·4
Flexural strength	MPa	400
Young's modulus	GPa	370
Akali resistance		Moderate
Acid resistance (except HF)		Excellent

TABLE 4
Geometric characteristics of a cross flow heat exchanger with rectangular channels

Characteristics	Unit	Value
Outer dimensions	mm^3	300 × 150 × 300
Volume total	cm^3	13 500
void fluid 1	cm^3	3 809
void fluid 2	cm^3	3 809
Total mass	kg	17·1
Hydraulic diameter		
fluid 1	mm	7·95
fluid 2	mm	7·95
Flow path length		
fluid 1	mm	150
fluid 2	mm	300
Geometric heat transfer surface	m^2	1·6
Specific heat transfer surface (related to total volume)	$m^2\ m^{-3}$	118·5
Specific surface weight	$kg\ m^{-2}$	10·69

First of all a single element was tested in pure cross stream. In the second test three elements were connected in the flue gas direction in line, for the air stream they were connected in parallel. For the third test the cross counter-stream connection was measured by installing reverse chambers on the air side. With a mass flow of 600 kg h^{-1} and a flue gas temperature of 750°C a single element achieves approx. 250°C air temperature in a pure cross-stream (Fig. 10). With three elements in a line for the flue gas and parallel for the air, the temperature can raise to 350°C, with an effective cross-stream connection of approx. 420°C. Increasing the flue gas temperature up to 1200°C, approx. 380°C, 610°C as well as 790°C are reached. This means an improvement of 88%.

The thermal efficiency is shown as the relation of maximum transmittable power to the actual transmitted power (Fig. 11). For 600 kg h^{-1} mass flow and 1200°C flue gas temperature the thermal efficiency for the single element is 28%, for line-parallel-connection 48% and for the cross counter-flow 68%. This reduction of the exhaust gas temperature leads to a clear decrease in efficiency.

The necessary blower power can be derived from the data for the pressure losses (Fig. 12). They diminish the total efficiency for a certain

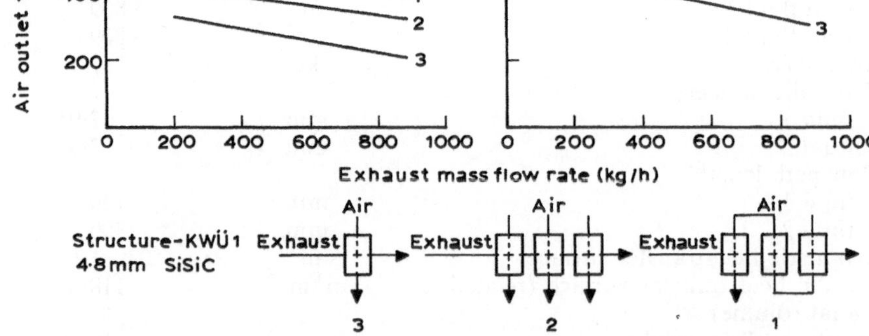

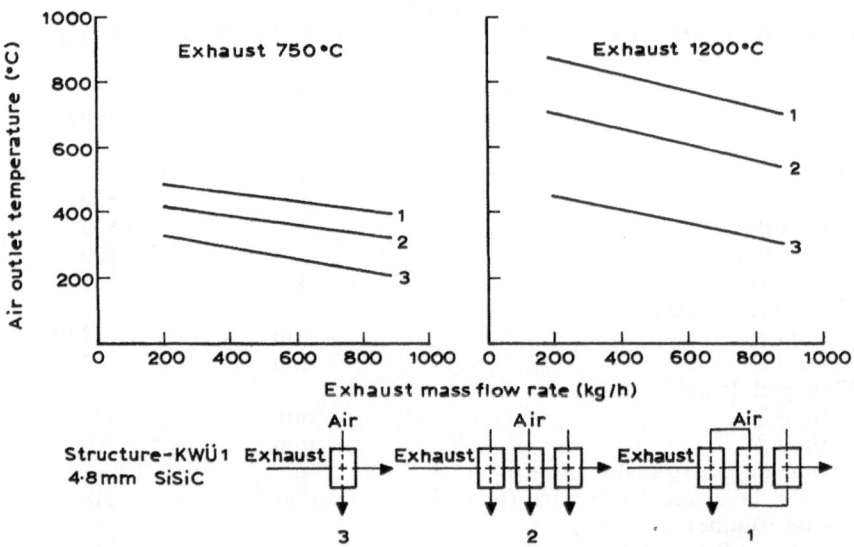

Fig. 10. Air temperature as a function of mass flow at different exhaust gas temperatures.

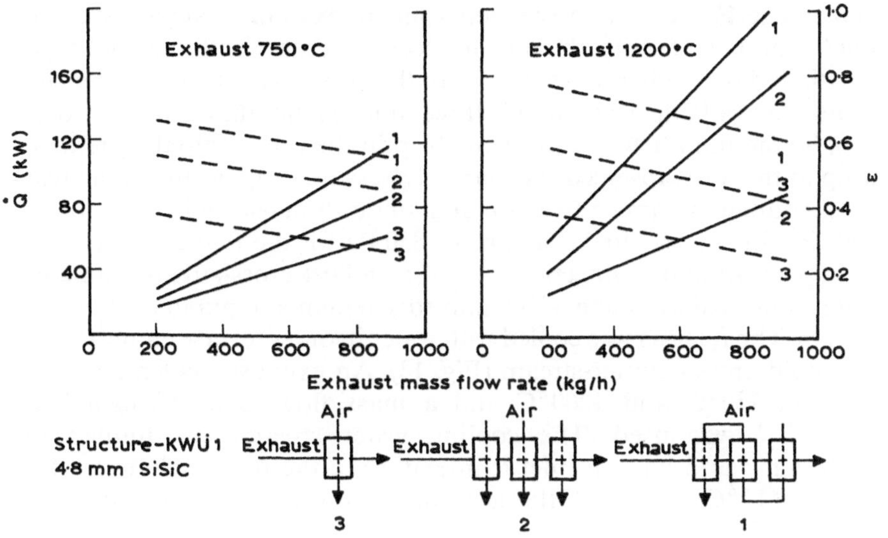

Fig. 11. Thermal efficiency and recuperated energy as a function of mass flow at different exhaust gas temperatures. (——) $\dot{Q}$, (- - -) ε.

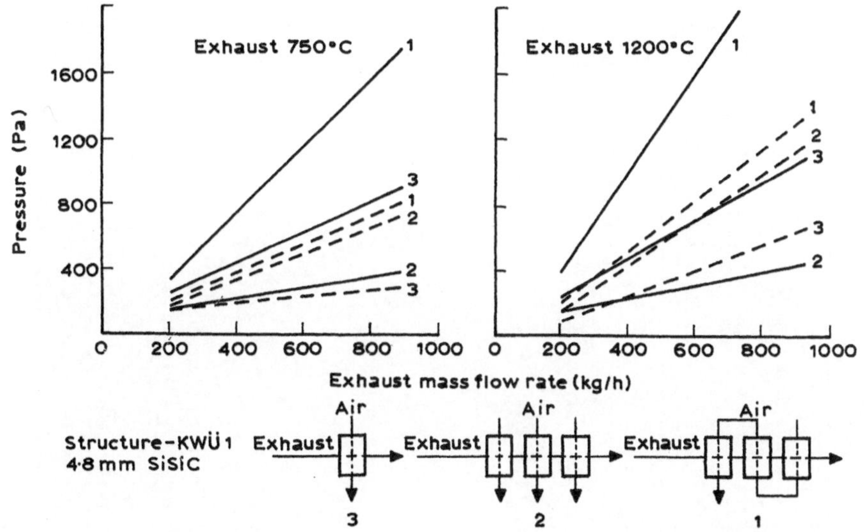

Fig. 12. Pressure drop as a function of mass flow at different exhaust gas temperatures (——) air, (- - -) exhaust.

percentage. However, it can be seen from this example that the losses are not too high. Naturally, the pressure losses increase with the total path length and the higher mass stream. In the cross-stream connection, the available available air section is three times higher than with the other two. So the ratio between pressure drop in the flue gas to the pressure drop in the air is reversed. The largest pressure drop occurs at the real cross counter-stream connection and amounts at 600 kg h^{-1} to approx. 1600 Pa for the air side and approx. 850 Pa for the flue gas side.

In a gas/liquid connection two of standard cross-stream modules were connected to a cube of 300 mm edge length and put to the flue gas stream. The flue gas was cooled with water in outer reverse chambers in a tenfold cross counter-stream (Fig. 13). An exhaust gas temperature between 250°C and 1200°C and a mass flow from 270 kg h^{-1} to 600 kg h^{-1} was used. The cooling water stream was limited to 2000 kg h^{-1} and had an inlet temperature to the heat exchanger of a constant 12°C. At a 250°C flue gas temperature and a low mass flow the

Fig. 13. Condensing heat exchanger unit with two standard elements.

condensing field is reached. One can conclude from those results that this unit can bring out approx. 200 kW power from a flue gas stream. During all tests this type of heat exchanger always remained at the temperature of the cooling water.

5. CONNECTING TECHNIQUE

The maximum size of the elements for the ceramic heat exchanger is limited due to the production techniques. Normally it is impossible to heat-seal or solder ceramic parts. This problem was solved by the module concept and a ceramic adapted joining and connecting technique. The open block shape of the heat exchanger elements with always the same outer dimensions, but variable inner structure, permits the modular construction of larger units with very different arrangements for industrial use.

Joining and sealing of the elements against each other and to the periphery is made in the direction of one product stream. In principle, the same techniques as used for constructing graphite heat exchangers can be applied. The sealing element for high temperatures for example can be a ceramic fibrous felt. When liquid and abrasive media are handled, graphite and PTFE seals, as well as glass solder combinations, can be used. Thereby the block shape offers sufficient space for reliable sealings between the blocks, and for the joining elements.

6. CERAMIC HEAT EXCHANGERS IN INDUSTRIAL USE

Since December 1986 in different porcelain kilns up to six single elements have been in service. Due to the high flue gas temperature of 1430 °C and the corrosive substances that escape from porcelain glaze, no metal heat exchangers can be used. The elements are used in the cross-stream system, each of them provides preheated combustion air of approx. 440 °C for two burners. Due to recuperation approx. 25% of the energy will be saved from a complete kiln run.

The example in Fig. 14 shows two heat exchanger units with each of two elements in cross counter-stream connection. The modules which are connected in parallel in the flue gas stream (1200 °C) of a high temperature tunnel kiln preheat combustion air for the kiln to approx. 700 °C.

Fig. 14. Two recuperator units with each two elements in cross counter-flow connection.

In a movable heat exchanger unit (Fig. 15) three elements are used. The flue gas flows linearly upwards through the three elements; the air is conducted through the three elements in the cross counter-stream system. The compact ceramic unit has a heat exchanging area of 4·35 m^2 with outer dimensions of 100 × 700 × 600 mm^3 and a weight of 280 kg. With this unit combustion air temperatures of 1100°C have already been achieved. A kiln manufacturing company already produces different apparatus with two and three elements.

A unit containing 15 ceramic heat exchanger elements (Fig. 16). has been built for cleaning air. The first level with nine elements preheats gas in three parallel tractions to approx. 500°C. A second level with six elements uses the remaining energy of the exhaust gas to produce 200°C hot air for a dryer.

An example of an exchanger in a chemical application is the acid cooler in Fig. 17. This acid cooler is used for cooling concentrated sulphuric acid at 250°C. Neither metals nor graphite can be used in such a case. Six elements are connected in a line on the acid side and are cooled by water or vapor in an enamelled steel container. The feeding pipes are also lined with corrosion resistant materials.

Fig. 15. Movable heat exchanger unit with three elements in cross counter-flow connection.

7. CONDENSING HEATING UNIT FOR DOMESTIC APPLICATION

The domestic boiler the schematic view of which is shown in Fig. 18 has been developed in coopeation with the Institut für Reaktorentwicklung of the KFA in Jülich, FRG. Besides the usual peripheral equipment of a boiler it consists of a gas blow burner,[6] a water cooled metallic combustion chamber,[7] a ceramic heat exchanger[9] and condensation separator.[10]

Depending on the interpretation the device can be used in switch on/switch off service, but can also work in a modulating mode. The hot gas is conducted downwards and cooled down by the heating water circulation from initially 1000 °C to an exhaust temperature which is approx. 5 °C above the heating water back flow temperature. The heating water is conducted in counter flow and extracts approx. 60% of the energy of the hot gas in the ceramic heat exchanger. The remainder is absorbed in the metallic combustion chamber. The combination of

Fig. 16. Heat exchanger unit with 15 elements for air cleaning.

an additional heat exchanger and a range boiler allows preheating of water for domestic use.

Figure 19 shows the heat exchanger without lining in side view. In Table 5 the important performance data of this heat exchanger are indicated. With these data the requirements of the German Standard according to DIN 4702/part b arc fulfilled.

8. SUMMARY

- A modular ceramic heat exchanger system has been developed. The core component is an open block element with rectangular or round channels.

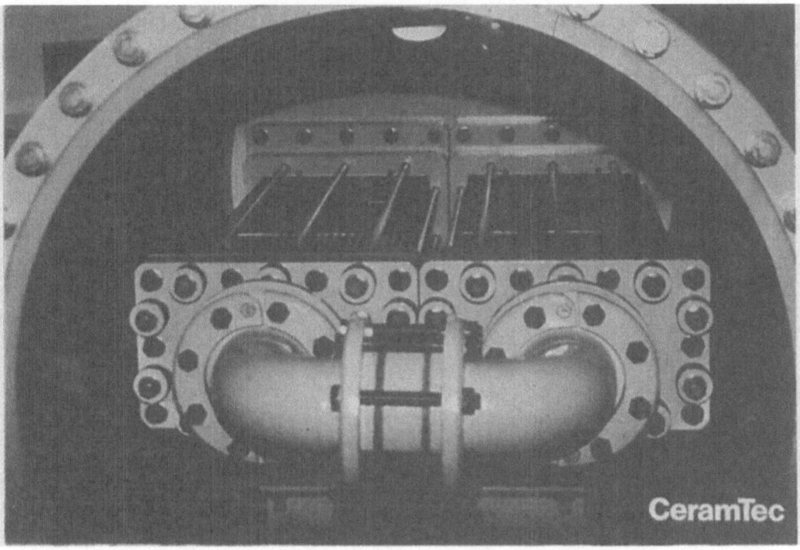

Fig. 17. Acid cooler for concentrated sulphuric acid.

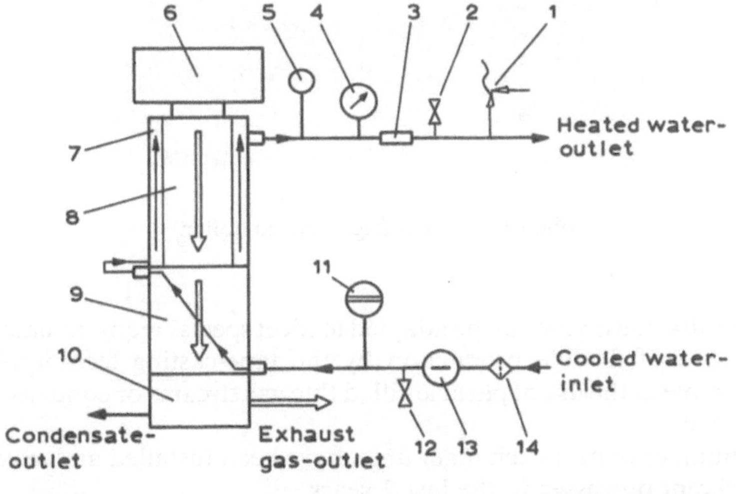

Fig. 18. Schematic view of the condensing ceramic boiler. 1. Safety valve, 2. evacuation, 3. water-flow-detector, 4. manometer for pressure, 5. thermometer, 6. gas-blowing burner, 7. watercooled-combustion chamber, 8. hot-gas-track, 9. ceramic heat-exchanger, 10. condensate trap, 11. expansion vessel, 12. draining tap, 13. pump, 14. filter.

Fig. 19. Condensing ceramic boiler.

- The block structure can be adapted to meet special requirements with the aid of flexible production by the tape-casting technique, for example in the use of particle-filled flue gas streams or condensation service.
- A number of heat exchanger units have been installed and tested in different processes in the last 2 years.
- By use of a ceramic counter flow heat exchanger a condensing boiler has been developed. Related to the net calorific value in condensing service the efficiency is 104%. After a test cycle of 3 years no corrosion has been observed in the heat exchanger.

TABLE 5
Performance data for the condensing ceramic boiler

Property		Unit	Value
Power range		kW	9–19
Pressure drop (total)			
gas side		mbar	0·6–0·7
Pressure drop			
gas side		mbar	0·1–0·2
water side		mbar	2
Combustion chamber temperature		°C	1 000
Temperature difference between flue gas exhaust and back flow water		°C	5
Condensation rate of H_2O vapour arising at combustion		%	>80
Efficiency related to the net calorific value		%	104
Flue gas emiision within the power range at excess air = 1,1			
	CO	ppm	33–65
	CO_2	ppm	10·7
	NO_x	ppm	40–45
pH-value condensation			2–3

REFERENCES

1. Foster, B. D. and Patton, J. W., *Advances in ceramics, Vol. 14,* American Ceramic Society Inc., Columbus, Ohio, 1985.
2. Williams, J. C., Doctor-blade process. In: *Treatise on materials science and technology. Vol. 9,* F. F. Y. Wang (ed.), Academic Press, New York, San Francisco, London, 1976, pp. 173–97.
3. Onada, G. Y. Jr, The rheology of organic binder solutions. In: *Ceramic processing before firing*, G. Y. Onada and L. L. Hench (eds), John Wiley, New York, Chichester, Brisbane, Toronto, 1978, pp. 236–51.
4. Mistler, R. E., Shanefield, D. J. and Bunk, R. B., Tape casting of ceramics. In: *Ceramic processing before firing*, G. Y. Onada and L. L. Hench (eds), John Wiley, New York, Chichester, Brisbane, Toronto, 1978, 411–48.
5. Richerson, D. W., *Modern ceramic engineering*, Marcel Dekker, New York, Basel, 1982.
6. Heinrich, J., Schelter, H., Schindler, S. and Krauth, A., Process for manufacturing heat exchangers from ceramic sheets. US Patent 4.526.635, 1985.
7. Ganz, R., Schelter, H. and Heinz, O., Wärmetauschermodul aus gebranntem keramischem Material. EP 0 274 694 (1989).
8. Förster, S., Quell, P., Heinrich, J., Huber, J. and Schelter, H., Ceramic residential

boiler with condensation of combustion water vapor. *Proc. Int. Symp. Condensing Heat Exchangers,* Battelle Columbus Laboratories, Columbus, Ohio, 1987.
9. Heinrich, J., Huber, J., Schelter, H., Ganz, R., Golly, R., Förster, S. and Quell, P., Compact ceramic heat exchangers: design, testing and fabrication, *Brit. Ceram. Trans. J.,* **86** (6) (1987) 170–205.

11

Nitride Bonded Carbide Engineered Ceramics

G. HOLLING

Thor Ceramics Ltd, PO Box 3, Stanford Street, Clydebank, G81 1RW, UK

1. INTRODUCTION

During research of the markets for advanced ceramics a range of industrial applications was identified, which required ceramic components with properties superior to those currently available. All of these processes were relatively new and essentially state-of-the-art in their particular applications in non-ferrous metallurgy and industrial process heating. While it was known that the properties of advanced ceramics were more than adequate to cope with the severe operating conditions experienced in these novel processes, such advanced ceramics were and still are only available in very limited shapes and sizes. They are also extremely expensive.

In the development of these applications the only option was to use ceramic components which were derived from coarse-grain refractory brick technology, and which were subject to short and unpredictable lives. This indicated that not only was material quality inadequate for the application, but also that control of the manufacturing process was insufficient. Occasionally components did give acceptable economic lives, albeit infrequently. This situation had restricted the growth of these new processes in many cases, although it should be emphasised that without the existence of these conventional refractory products in the shapes and sizes required, the development programmes could not have been considered for commercial exploitation. In several cases advanced ceramics had been successfully used in small-scale prototypes, but scaling-up of the application extended the size of the ceramic components to greater than those available in the existing range of advanced ceramics.

In reviewing the situation it became patently obvious that the development of new materials suitable for advanced applications would require massive investment over many years, with no guarantee of success. With the certain knowledge that many large companies had for many years been working on such projects as the applications of ceramics to engine components with little commercial success, it was clear that Thor were unable to compete in this highly competitive area.

The properties of carbides and nitrides and, in particular, silicon nitride bonded silicon carbide, made them first choices for many of the new applications. The form in which this material is most commonly available is a mixture of silicon nitride, silicon oxynitride and silicon aluminium oxynitride phases, together bonding to the coarse silicon

TABLE 1
Properties of silicon nitride bonded carbide

Apparent porosity	15–17%
Bulk density	2·55–2·65 Mg m^{-3}
Apparent solid density	3·01 Mg m^{-3}
Thermal conductivity	16 W m^{-1} K^{-1}
Coefficient of expansion	4·6 M K^{-1}
Bond strength @ 20°C	45 MPa
@ 1300°C	40 MPa

Chemical analysis

SiC	75%	Al_2O_3	1·3%
Si_3N	22%	Balance	1·0%
SiO_2	0·7%		

Thermal shock
Very good due to tough strong bond and high thermal conductivity

High temperature load bearing
Excellent. Up to 1700°C in protective atmospheres

Abrasion resistance
Extremely good especially to liquid and airborne particulate solids

Corrosion resistance
Highly resistant to most chemicals and liquids at low temperatures, molten non-ferrous metals and their slags

carbide grains. Nitride bonded carbides are available from about ten commercial suppliers throughout the world. Closer examination showed us that despite its humble refractory background, its general properties over a wide range of conditions and environments are superior to most other materials available as large and complex shapes. Table 1 lists these properties. Unfortunately, a conventional table of properties based on results of standard testing methods tells only a limited story. It is only when a large pool of operating experience of generally imprecise nature is gained, that a fuller and more accurate perception of a material's performance capabilities emerge.

It was decided at Thor that a range of engineered ceramic compositions could be developed by an 'Incremental programme' working away from the traditional coarse refractory compositions and at the same time tailoring each mix and manufacturing procedure to the requirements of the product's service environment. Working in this manner a range of engineered ceramics has been established for applications including:

— high temperature radiant heater tubes,
— immersed heater tubes for non-ferrous metal heating,
— burner quarl inserts for regenerative burners,
— curtain wall baffles for reheating furnaces,
— low and high pressure die casting feeder tubes.

2. THE INCREMENTAL APPROACH

Thor had 'in depth' experience amongst its senior personnel in silicon carbide refractories, and also a wide range of more specialised refractories and fine grained ceramics. It was believed that by careful manipulation of existing silicon nitride bonded silicon carbide recipes and granulometry it would be possible to improve upon the existing properties. It was also believed that it would be possible to maintain the properties achieved in simple brick shapes for much more complex shapes by improvements in manufacturing techniques.

It has been a long tradition in the refractories world to quote standard brick properties for a class of materials with a disclaimer for special shapes knowing full well that the properties of the latter were not only well down on standard brick properties as published but varied considerably throughout the piece. Whilst Thor had extensive fabrication technology and experience, its speciality was in the field of isostatic pressing utilising a range of press vessels of varying sizes and operating pressures, and it was in the isostatic fabrication area where the largest increment of improvement was anticipated, and has actually been achieved. It has been possible not only to maintain standard brick properties in complex thin wall tubes but also to improve significantly on these properties, and also their consistency within an individual component by judicial use of isostatic compaction techniques and advanced tooling technology. A diagram of a large isopress is shown in Fig. 1.

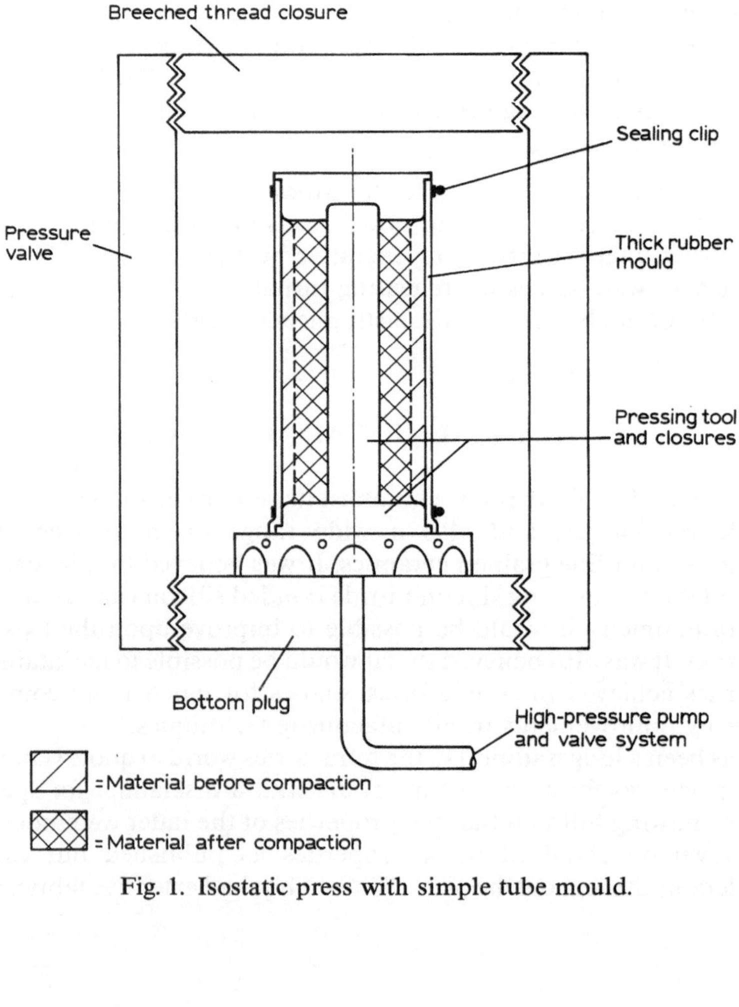

Fig. 1. Isostatic press with simple tube mould.

Last but not least, the nitriding process was seen as a potentially weak link. The acquisition of a high integrity controlled atmosphere furnace from AEA at Harwell meant that, if required, ultra high vacuum, or several bars pressure, could be applied when nitriding shapes up to 2 m in length and in excess of 1 m diameter. The quality of nitridation could be guaranteed to be as good as the best silicon nitride products, and superior to and more consistent than products from existing nitride bonded carbide manufacturers. A diagram of the nitriding furnace is shown in Fig. 2 together with a description.

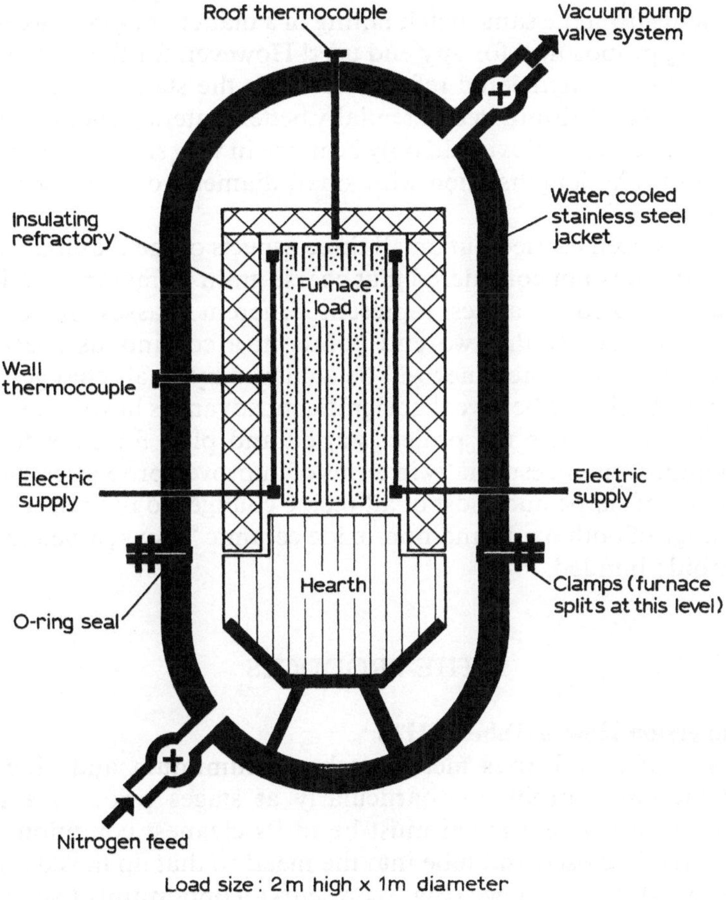

Fig. 2. High integrity nitriding furnace.

Each of the increments of improvement in the recipe and formulation area, fabrication process and nitriding process, in themselves may not have been individually significant. Cumulatively, however, they were very significant, and at least two out of three of these benefits have also been attained in products made by the uniaxial pressing and casting, fabrication routes. One of the major advantages superimposed on the uprated properties has been the great benefit obtained by the increase in reliability, and in consistency of properties.

In many of the subsequent case studies the major disadvantage of previously available products was their unpredictability of life. Sometimes components would last 6 months with apparently identical components from the same batch failing in a matter of days or weeks. A frightening proposition for any end user! However, for the end user in most cases such engineered refractories were the state-of-the-art, and size requirements eliminated potentially better materials, because most other advanced ceramics could only be made in very small sizes, or if in the form of long lengths, then with small diameter or cross-sectional limitations.

Work has been carried out on all three aspects of the manufacturing process, and it is not considered that coarse grain refractories and fine grained advanced ceramics represent different classes of ceramic material, but merely the two end points in a continuous spectrum. Moving from one to the other will be achieved by small changes on a continuous basis. We believe that significant advances have been made in the 3–4 years since the project started and plan for considerable further improvement especially in terms of improved properties, greater complexity of shape and most of all recipe changes to include a much wider range of both oxide and non-oxide ceramic 'alloys', albeit nitride and carbide bonded.

3. THE PRODUCTS

3.1 Immersion Heater Tubes (IHT)

Heating of non-ferrous metals such as aluminium and zinc has always presented problems, particularly at stages prior to a final forming process where metal must be in its cleanest condition. The immersion of a closed end tube into the metal so that up to 900 mm of the tube is below metal level (Fig. 3) presents an opportunity to conduct heat from a power source inside the immersed tube through its wall into

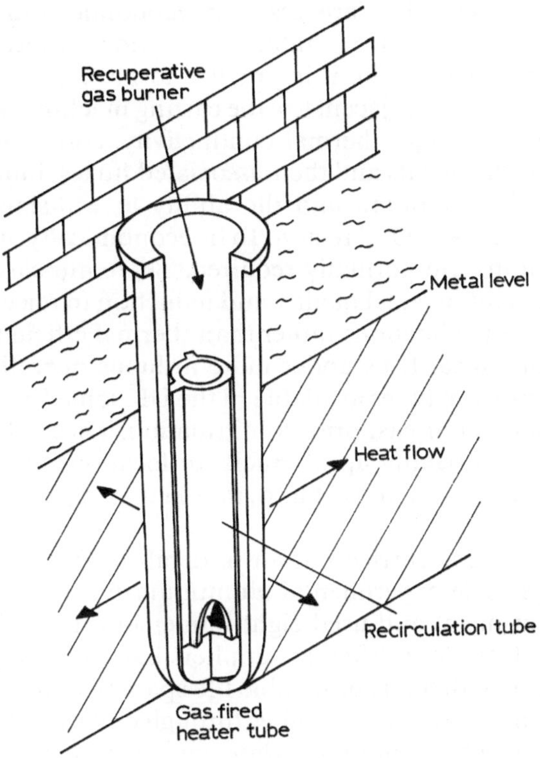

Fig. 3. Immersion heater tube gas fired.

the metal. Because the inside of the tube isolates the source of the heat from the metal, products of combustion cannot make contact and contaminate the molten metal, nor can they agitate or oxidise the metal. Thus purity can be maintained with a relatively rapid and efficient heat input. The purification process applied to molten aluminium immediately prior to casting at smelters, or after remelting of scrap at a rolling or extrusion mill, nearly always involves a heat loss. There has been a need to put heat into the system at this point. Purification systems usually involve a gas purge filtration and the addition of fluxes or chlorine in small controlled quantities. These processes normally take place in what is called an SNIF unit, or filler box, which is an independent structure. In such units the immersion heater offers the potential for heat input, without risk of detrimental effect from the heat source.

Immersion heater tubes are also an economic and convenient method for maintaining temperature in a wide variety of holding furnaces at die casters where it is practice to have a relatively small holding furnace located adjacent to a die casting machine. The required tube properties were high thermal conductivity, corrosion resistance against non-ferrous metals and their associated fluxes, imperviousness to molten metal penetration and the ability to withstand oxidising conditions in the 900–1200 °C region. To be economically viable lives in excess of 3 months are normally required. There are cases, however, where improvement in metal quality and reduction in reject levels have been the major considerations rather than thermal efficiency allowing for shorter economical lives under more arduous operating environments. The immersion tubes available in the UK some 5 years ago were either hand rammed or cast prior to nitridation. These often failed in service due to the opening up of cracks in locations associated with laminations inherent in most hand rammed components, or from casting defects.

In 1987 British Gas carried out an in-depth independent study on tubes in a large scale experimental aluminium furnace. This furnace melted 1 t h^{-1} of metal and used eight immersion tubes. The existing available tubes from four different suppliers had an average life of 37 days but with no predictable or reliable life span. Failure was normally due to metal penetration into the tubes through cracks and could occur minutes, days or weeks into the campaign. In trials with isopressed carbon bonded silicon carbide tubes, failure occurred in preheat, probably due to oxidation which limits this potentially very refractory bonding system to low melting point metals such as zinc. It was apparent that the main requirement of the end user was a more predictable and reliable tube life, preferably as long as possible, but with a 3 month minimum economic target. It was obvious to us that the highly varied life was due to fabrication defects and not essentially the material. It also became apparent in the early days of our work that this product was an ideal target for our introduction to the market. It was decided therefore that the material 'class', that is nitride bonded carbide, was correct but the fabricating process then employed — casting and hand ramming, were totally unsatisfactory due to the latent defects introduced by drying and water concentration gradients, and by the density/layering defects of hand ramming.

During the latter half of this test period Thor tubes were introduced to the market place. The results of the British Gas tests are unequivocally stated in a paper published in 1988. The isopressed Nicarb tubes made by Thor increased the average immersion tube life from 37 days but with no predictable life span to 270 days with typical lives of 6–10 months. The main parameter that we aimed for was a 'defect free' end product. At that point in time it was only required that the material itself would not fail before 3 months, the target minimum life.

In order to achieve this defect-free end product, the recipe was developed from traditional compositions as a compromise between maximum corrosion resistance as exhibited in conventional refractory grade materials and a slightly more open textural property advantageous in respect of thermal resistance. The best of the existing materials had apparent porosities of 15–19%, often within the same piece depending on fabrication method. Thor's recipe yielded an end product of 20–22% apparent porosity. The overall chemical analysis differed only slightly, and the mineralogy was much the same, although Nicarb contained little or no oxynitrides compared to competitive products. Quantitative chemical analyses of nitrogen ceramics is notoriously difficult and the mineralogical analysis is even less precise, particularly when dealing with several similar components in the presence of silicon carbide and residual silicon and Carbon.

The results of the independent trials showed that the benefits of achieving an isopressed homogeneous body more than compensate for the slight increase in apparent porosity, the whole component being free from flaws which could be crack initiators when subjected to thermal cycling in service. As part of the on-going development programme subsequent manipulation of the recipe has enabled the porosity to be reduced to $<13\%$ with little change in mineralogy and still maintaining 'fabrication friendly' mixes. This improved tube is now entering the market and expected to give even longer service lives particularly in hostile environments. Other avenues of material development are under investigation and will be discussed later.

3.2 Radiant Heater Tubes (RHT)

The success of the isopressed immersion tubes led to a request to consider manufacturing other tubular products which were suffering from poor performances assocated with inferior compaction and shaping processes. The first of these was a tube for use in industrial radiant heaters to provide a controlled or quiescent furnace atmosphere. The operating temperatures of these tubes can exceed 1250 °C internally and they require sufficient mechanical strength at this operating temperature to resist deformation by creep. These radiant tubes were required in 2 m lengths, with internal diameter of 150 mm, and 15 mm wall thickness. The first immersion heater tubes had been less than 1 m in length and had a much thicker (25 mm) wall. This new tube length was twice the length previously made and approximately $\frac{2}{3}$ the wall thickness. As the difficulty of handling tubular shapes is approximately related to the wall thickness to length ratio these new tube sizes were believed to be approaching the limits of existing fabricaton technology.

At the higher operating temperatures it was also recognised that mechanical and oxidation resistance properties would be improved by reducing the porosity of the material. The development work already carried out on the batch properties for immersion heater tubes to reduce the apparent porosity while maintaining ease of manufacture, was readily transferred to the manufacture of these radiant tubes. When combined with improvements in isostatic compaction tooling and handling equipment, the net result was that since the radiant tube project was started not a single case of premature failure has been reported. The initial target lives of 6 months are now being easily exceeded, with the original test bank of radiant heater tubes still in service with a realistic service life of 2 years being targeted. Typical bulk properties of these radiant heater tubes are given in Table 2.

It can be seen from Table 2 that the porosity of the isopressed material has been reduced from ~20% to ~14%. This is slightly better than a typical refractory pressed brick. The use of isopressing has again allowed the fabricaton of defect-free pieces.

The success of the 2 m radiant heater tubes in simulative cyclic testing convinced the inventor of the system to fund a short term R & D programme aimed at developing the next generation of materials for this particular radiant heater tube application. The previous supplies of tubes had given varying lives, some as little as 3 weeks and Thor's isopressed quality tubes were anticipated to give lives of >3 months

TABLE 2
Typical bulk properties of early and latest high density radiant heater tubes

	Latest tubes	Early tubes
Apparent porosity	13–15%	15–20%
Bulk density	2·58–2·68 Mg m^{-3}	2·53–2·63 Mg m^{-3}
Apparent solid density	3·01 Mg m^{-3}	3·01 Mg m^{-3}

which was about the minimum economically acceptable. When this R & D programme was eventually started the isopressed tubes had been in service approximately 5 months and showed, with one exception, no expansion in length, the main parameter used to assess oxidation. This exception was in a low temperature part of the kiln where much more severe problems had been occurring with hand rammed tubes from the original source. It was realised that as a relatively quick feedback was required from this R & D programme then the programme must be kept simple and it was agreed with the client to limit receipe changes and concentrate on reducing the porosity of the finished product and improve its oxidation resistance. The results of this short R & D programme was as anticipated. A small but very significant improvement had been made by reducing the typical porosity to about 12% whilst still keeping the, essentially original, user friendly recipe.

It became apparent during the course of this R & D programme that the average radiant tube life in a 24 tube tunnel kiln was going to be well in excess of the hoped for 6 months, and that the kiln operators were thinking of lives of up to 2 years as being possible and currently being aimed for at the time of writing this paper. This along with the original tube properties, the revised mixes and processes, offers still further potential.

Experience being gained from these two applications and from other new applications to be described later, indicates that for short operational periods it may be possible to increase the internal operational temperature of NICARB radiant heater tubes to 1650 °C, and if required reduce the wall thickness. The implications of being able to produce ceramic radiant heater tubes to provide working temperatures in the furnaces of up to 1600 °C are substantial.

Ni–Cr alloy radiant heater tubes have been widely used for many years but are limited in general to below 1000 °C in service. Their general shape and design has been governed by the metal's ability to be produced with 3 mm thick walls, in complex shapes including threaded fixings and advanced alloy welding techniques. Ceramic materials have inherent limitations of design which means, unfortunately, that there is limited scope for direct conversion or by replacement ('retrofit') of high temp alloys by cheaper and longer lived ceramics. The nitride bonded carbides described could easily be used for these applications and outperform them in most cases, but for the unfortunate design problems associated with retrospective fitting. As manufacturing techniques improve, some but not all of these problems may be overcome.

3.3 Alternative Thermal Applications

During the radiant and immersion heater trials it become obvious that the Nicarb materials in isopressed form have excellent thermo-mechanical properties. However, the determination of operational limits was difficult to quantify and prohibitively expensive to establish by simulative testing. Concurrent with the burner tube developments, and as a result of information from the 'industrial grapevine' we were approached by two customers who had problems with tubular form components for use under high thermomechanical stresses. Both of these were apparently suitable for isostatic press formation in Nicarb materials and the customers were willing to conduct trials with us.

3.3.1 Burner quarls

A new high efficiency, regenerative burner system was experiencing failures in the burner quarls made from conventional refractory materials. The operating conditions for these quarls were extremely severe, with thermal cycling from 1480 °C to 1250 °C every 70 s; existing materials could not cope with the resultant thermal stresses. Quarls made from isopressed Nicarb were manufactured for trial and have subsequently given satisfactory performance in this most arduous operation giving service lives of 1 year when operating on a variety of fuel types.

3.3.2 Thin wall complex shapes

A furnace manuacturer was experiencing difficulty with refractory components to make a curtain wall in a rotary hearth furnace operating at 1350 °C. This curtain wall separates two zones of the furnace and allows precise control of atmosphere and temperature in each zone. The curtain wall was constructed from a bank of tongue-and-grooved tubes suspended by pins from an overhead support rail. The existing refractory materials were made by an extrusion process and suffered from warpage which made initial installation difficult. In operation this deformation continued reducing the efficiency of the curtain wall and making it impossible to replace a tube if one became damaged during the furnace campaign. Figure 4 shows a diagram of the curtain wall.

The experience gained in the manufacture of the 2 m radiant tube indicated that components with wall thicknesses less than 15 mm could be isopressed, but that problems of dimensional accuracy increased with increasing component length. As machining of thin wall components made from coarse grained materials is difficult, and results

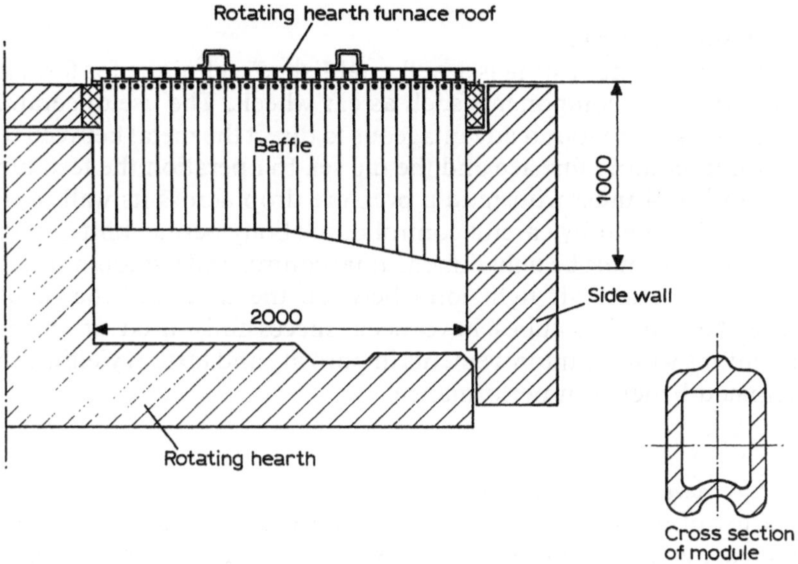

Fig. 4. Suspended curtain wall (modular).

in potentially high wastage, it was obviously desirable to fabricate to near net shape and, therefore, essential to use a tolerant mix composition plus specially designed tooling for the subsequent stripping and handling processes.

Once these process developments were made it has proved possible to press the curtain wall tubes at lengths of 1·5 m with wall thicknesses down to 6 mm, despite the complex external and internal profile. Wall thickness has been controlled to less than ±1 mm and no significant warpage occurs. Several curtain wall systems have now been supplied with service lives in excess of 1 year being achieved.

3.4 Non-Ferrous Metallurgical Applications

Silicon nitride bonded silicon carbide materials offer potentially advantageous properties with respect to thermal shock resistance, non-wetting by aluminium and its alloys and corrosion resistance against both metals and their associated slags. The use of isopress forming techniques allows tubular components to be made with high integrity and to precise dimensions. Several potential applications have been investigated.

3.4.1 Pressure feeder tubes

Low pressure die casting is a well established technique for forming aluminium alloy components such as car wheels. The use of ceramic feed tubes has been found advantageous to direct the metal between the pressurised holding furnace and the die set. In operation the feed tube shown in Fig. 4 must withstand a pressure drop across its wall in the order of 0·7–1·5 bar by careful control of the body permeability. At the collar end there must be tight dimensional control and a smooth surface to ensure a perfect air tight joint between the tube and the mould surface. Nicarb riser tubes have been successfully used at various installations with the improvements in porosity and texture yielding the anticipated benefits in performance.

3.4.2 Electromagnetic pump bodies

The use of electromagnetic pumps to move molten aluminium alloys is increasing in popularity. Various systems are in operation, all utilising a ceramic tube or canal immersed in the metal bath. One pump tube is shown in Fig. 5. The design demands a high dimensional accuracy over its metre length and the rectangular slit must be tightly controlled to better than 1 mm duration over the tube length. Nicarb materials and isopressing were successfully used to produce these tubes but the external surface tolerances required that the pieces be diamond ground. Various trials resulted in the adoption of a two-stage finishing technique involving machining followed by diamond grinding in the unnitrided form. Unfortunately problems have been experienced with this tube form, mainly associated with the clamping systems needed to maintain the seals required to resist the overpressure of the internal aluminium. Further design modifications are under consideration to reduce these problems. Additionally development trials are underway to modify the material recipe to facilitate easier machining and an improved surface finish.

Work carried out some years ago on silicon nitride has also been continued for producing large size feeder tubes up to 2 m in length. This material is fine grained and when partially nitrided machines well. Reaction bonded silicon nitride is, however, considered only as a starting point for an incremental approach to improve its properties, particularly thermal shock oxidation resistance, and wettability by molten metals.

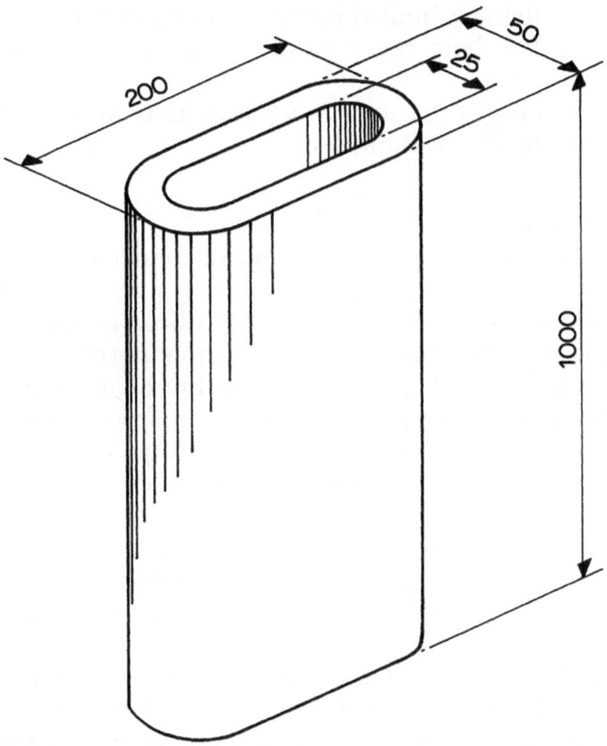

Fig. 5. Typical electromagnetic pump canal liner material: Si_3N_4 bonded SiC.

3.4.3 Size and thickness limitations of isopressed shape

There is obviously a limit to wall thickness when isopressing coarse grained materials and this combined with shape limitations will restrict its uses. In the absence of extensive machining it is not envisaged that wall thicknesses of less than 5 mm on a 2 m × 150 mm tube will be commercially available in the next 5 years. In the case of nitride bonded carbides the machining costs would be very high but fabrication methods involving the machining of partially nitrided isopressed silicon metal tubes and subsequent in-situ formation of silicon carbide are under consideration.

At present there is a theoretical limitation on the maximum diameter of less than 1 m, the largest press within the group, but there has been no commercial demand for diameters larger than 0·5 m. The size of the largest piece available is limited by the nitriding furnace although this could be extended from the present 2 m up to 3 m if there were commercial incentives to do so. Isopressing is not limited to simple regular shapes. It is possible to press a wide variety shell and box shape using advanced tooling techniques.

3.4.4 Material development

Currently only the basic silicon nitride bonded silicon carbide material is in volume production, but there is no reason why most refractory oxides should not be bonded with silicon nitride, although in our experience we have found the dominant bonding phase to be silicon oxynitride as a result of interaction with the oxide. Indeed nitride bonded alumina components are in use on a specialist steel casting machine, competing with conventional mullite bonded refractories and performing well giving multiple life operation in place of the anticipated single life.

If one considers the permutation of all the refractory oxides, nitrides, carbides, borides and silicides, then there is almost infinite scope for development work, most of which would break new technical ground. However, in our case development work is commercially driven and required to have a relatively short-term pay-back period. It is because of this philosophy that an incremental approach to development has been adopted. It is far better to seek a small positive beneficial improvement achieved in a short time scale, than to attempt quantum leaps in technology over much longer periods, with attendant higher risks of failure and no commercial payback.

The emphasis in our development programme stems from three requirements:

1. Increase in oxidation resistance in the region 800–1000 °C, and above 1200 °C. This is being achieved by reducing the open porosity and by incorporation of oxidation barriers in the form of internal glazes. Significant improvements have been made in the 800–1000 °C region but for current products such as radiant heater tubes life expectation has not yet been established and may well not be available for another year, some 3 years after product testing began!
2. Improvement of non-wetting properties towards non-ferrous metals by changing the chemistry with additions to the silicon carbide body of boron nitride, aluminium nitride and zirconia. Textural changes are also envisaged to minimise adhesion due to

mechanical coupling to surface defects. Though in its early stages, results from initial trials are encouraging.

3. Better machinability, to be achieved by textural changes such as reduction in grain size of the coarse silicon carbide fractions, and, what at first appears a drastic change, the development of what is essentially reaction bonded silicon nitride. Once again one is thinking of the incremental approach of silicon nitride bodies containing relatively small quantities of fine boron nitride, aluminium nitride and fine silicon carbide. Permutations of all four of these components are under investigation.

This work, still in its early stages, is primarily aimed at the non-ferrous metal handling industries. It is envisaged that within 5 years a highly oxidation resistant silicon nitride bonded carbide will be available for use in indirect heating applications up to 1500 °C with long and hence economic lives. In the field of non-ferrous metal handling it is envisaged that within 2 years a range of new fine grain reaction bonded silicon nitride will be available with better oxidation resistance than reaction bonded silicon nitride and improved corrosion resistance and non-wetting properties.

4. CONCLUSION

The incremental approach to the development of conventional refractory compositions has allowed a progressive improvement of a range of both materials and components to suit specific and often changing market conditions. Material compositions tailored to suit the particular operating conditions, and developments in isopressing technology, have resulted in the production of stress-free components of considerably greater size and complexity than previously available.

Thor components now realistically offer engineered ceramic solutions to operational problems at competitive prices. The extension of such applications will continue and accelerate provided the close co-operative procedures already esablished with existing customers are maintained with the ceramic specialist becoming involved with the design engineer early in the development phase to ensure that the unavoidable constraints relating to ceramic component design are designed out before they can become a limiting factor to the operational potential.

12

Progress with Zirconia Ceramics

I. BIRKBY AND H. HODGSON

Dynamic–Ceramic Ltd, Bournes Bank, Burslem, Stoke-on-Trent, Staffordshire, ST6 3DW, UK

1. INTRODUCTION

'Ceramic Steel?'[1] is the title of the first scientific paper to highlight the possibilities offered by the 'transformation toughening' mechanism which occurs in certain zirconia ceramics. Since the publication of this seminal work in 1975, considerable research, development, and marketing effort has been expended on this single material which offers the traditional ceramic benefits of hardness, wear resistance and corrosion resistance, without the characteristic ceramic property of absolute brittleness. The aim of this review is to select some of the more significant milestones which have measured the progress and problems of this fascinating material throughout its development as a structural ceramic. Particular emphasis has been placed on tetragonal zirconia polycrystals (TZPs), which represent (in terms of the number of producers) the most common form of zirconia in engineering applications.

The reader who requires a more in-depth analysis of zirconia in all its forms, should turn to an excellent reference work by Stevens,[2] or any of the proceedings of the three international conferences which have dealt exclusively with zirconia.[3-5]

Throughout this review and in other works on zirconia ceramics the usual nomenclature used to describe zirconia alloys is detailed below:

TZP tetragonal zirconia polycrystals
PSZ partially stabilised zirconia
FSZ fully stabilised zirconia

TTC transformation toughened ceramics
ZTA zirconia toughened alumina
TTZ transformation toughened zirconia

The common notation used in TZP literature involves placing the cation symbol of the stabilising oxide before the TZP or PSZ abbreviation. In some cases the amount expressed as mol % of the stabilising oxide will be indicated by a number before the cation symbol, e.g. zirconia containing 3 mol% yttria will be denoted as 3Y–TZP. Symbols corresponding to non-stabilising additions are placed behind the abbreviation. These additions are given as weight percentages, e.g. 3 mol% yttria - zirconia with 20 wt% alumina = (3Y–TZP)20A.

2. TRANSFORMATION TOUGHENING

Pure zirconia can exist in three crystallographic forms, cubic, tetragonal and monoclinic. All of these phases are variants on the cubic fluorite structure. The range of stability for each of the polymorphs is shown below:

$$\text{monoclinic} \overset{1170\,^{\circ}\text{C}}{\rightleftharpoons} \text{tetragonal} \overset{2370\,^{\circ}\text{C}}{\rightleftharpoons} \text{cubic}$$

The tetragonal to monoclinic transformation is of the greatest technological significance, due to the martensitic nature of the reaction and the accompanying 3–5% volume expansion. The martensitic reaction involves a diffusionless shape deformation of the tetragonal lattice by a shear mechanism. This shear mechanism causes any plane surface whether internal or external to become serrated on a microscopic scale. The large volume and shape deformations which occur on transformation cannot be relieved by diffusion. Instead they are accommodated by elastic or plastic deformation of the surrounding matrix. Martensitic reactions are usually athermal, that is, they occur only when the temperature is changing. This behaviour can be explained in terms of the increase in strain energy occurring on transformation opposing further transformation, and consequently the chemical driving force for the reaction having to be increased. Generally this is accomplished by undercooling below the temperature defining the start of the martensitic reaction (Ms).

The mechanism of stress-induced transformation toughening which is operative in TZPs, has been explored extensively by Lange.[6] His work quantified the thermodynamic criteria which must be satisfied for the retention of the metastable tetragonal zirconia, i.e. stabilising oxide content and distribution, particle size and the elastic modulus of the matrix. The retention of tetragonal zirconia is an essential prerequisite for toughening, as in the presence of a propagating crack the tetragonal particles are induced to transform to the monoclinic phase, with an accompanying volume expansion of 3–5%. Consequently, an advancing crack is subjected to a compressive stress, its progress is either halted or retarded, and the material is 'toughened' (Fig. 1). Figure 2 illustrates tetragonal precipitates which have transformed to the monoclinic structure within the cubic matrix of a Mg–PSZ. Similar behaviour is shown in Fig. 3 which shows the extensive transformation which has occurred in a Y–TZP which has a large grain size, and is consequently highly transformable. It is interesting to note the well-defined monoclinic twins which have developed in this material. In addition to toughening, providing the size and distribution of any inherent flaws are minimised, the measured strength of the material will also increase.

Significant progress has been made in terms of our ability to understand and predict the behaviour of transformation toughened ceramics. In turn this has led to a wide range of zirconia ceramic alloys

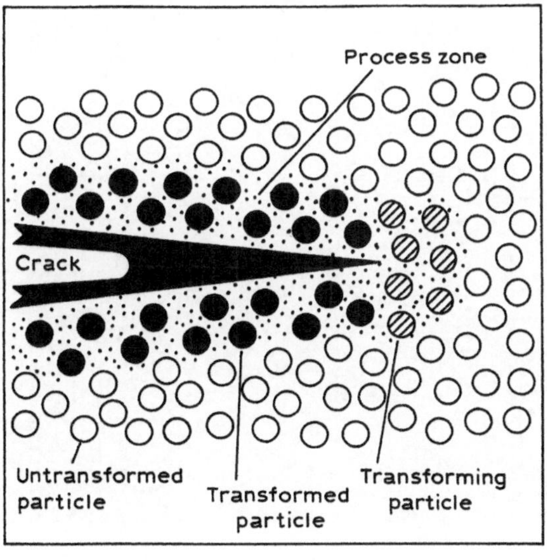

Fig. 1. Diagram illustrating the transformation toughening mechanism which operates in zirconia ceramics.

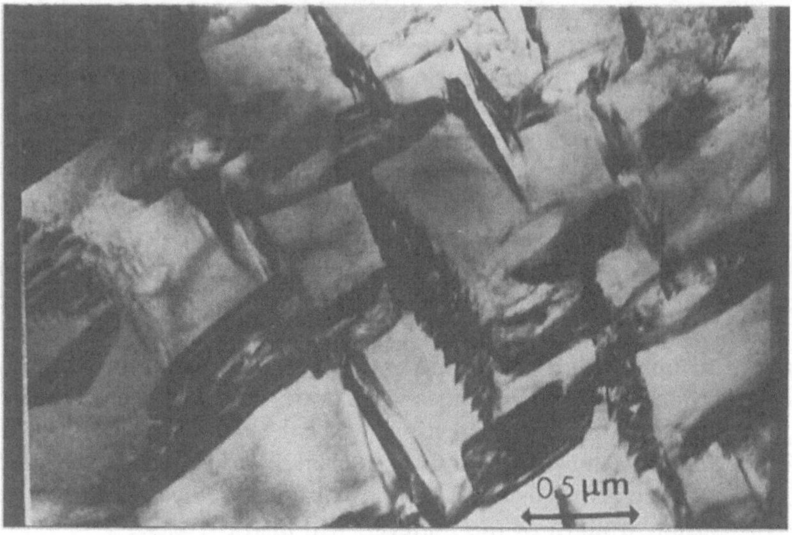

Fig. 2. Transformed tetragonal precipitates (Mg–PSZ).

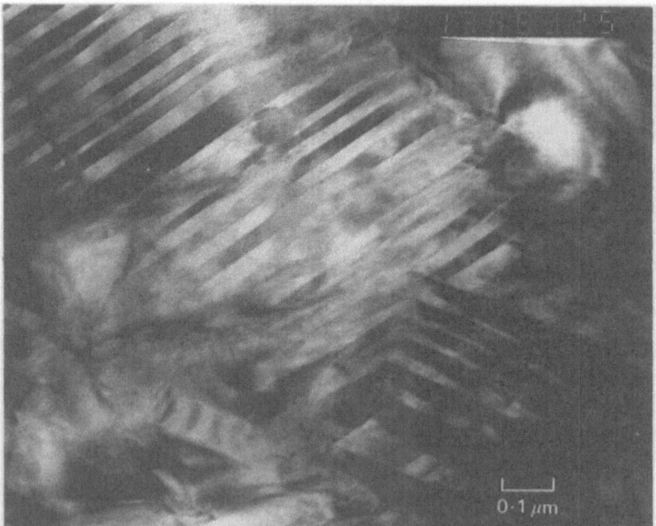

Fig. 3. TEM of twinned TZP (magnification ×74 000).

based on different stabilising additives and specifically tailored microstructures (Figs 4a–4c). The TZPs discussed in this paper can also be classed as partially stabilised zirconias (PSZs), as the metastable tetragonal phase is retained at room temperature, and can be induced to transform to the monoclinic phase (full stabilisation permits the retention of the high temperature cubic phase).

The common stabilising additives are usually rare earth oxides or similar compounds such as yttria, ceria, magnesia or calcia, which form solid solutions in zirconia and stabilise the high temperature phases. The amount of stabilising additive required to produce partial stabilisation is determined from the relevant phase diagram. The phase diagram for Y–TZP is detailed in Fig. 5.

3. RAW MATERIAL DEVELOPMENTS

Raw material suppliers have developed and are now offering a range of starting powders which are highly sinter-active. Their small particle sizes (25 nm) and high surface areas, together with high purity and chemical homogeneity[7] (Fig. 6a, 6b), allow complete densification at temperatures as low as 1400°C. In the early 1980s the dominant force in powder development was Japan, with its high purity chemically produced powders. Now, as we enter the 1990s, high quality powder production is being conducted in several countries (UK, France, Germany, Canada, Australia and others). Significant developments in powder production include the production of electro-fused zirconias with properties similar to the inherently more expensive chemically produced grades.[8] Other producers have considered the use of plasma synthesis and pigment coating technologies to develop novel powder morphologies with improved ageing properties (see Section 5)[9] (Figs 7a, 7b). Powders with optimised forming characteristics such as a high green density to prevent warpage are also now readily available.[10]

The increase in the competitive activity detailed above has led to significant decreases in powder costs (>50% in some cases), which in turn leads to a broadening of the potential application horizon for finished products. It should be remembered, however, that in most components the powder cost probably only constitutes 10% of the overall cost, as diamond grinding to precise tolerances often accounts for the major share of the manufacturing cost.

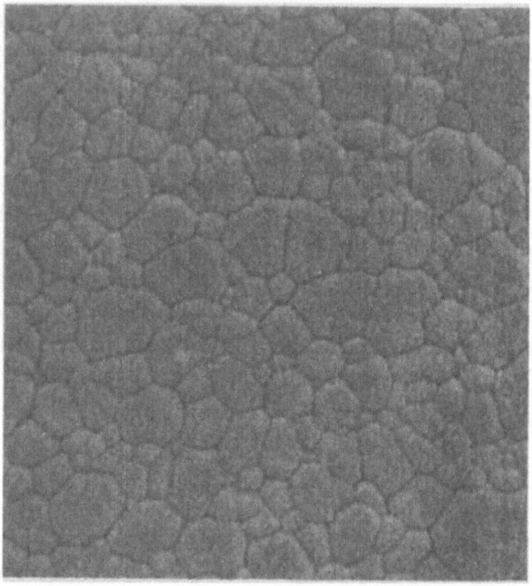

Fig. 4a. SEM of Y–TZP illustrating uniform grain size (magnification ×20 000), average grain size 0·5 μm.

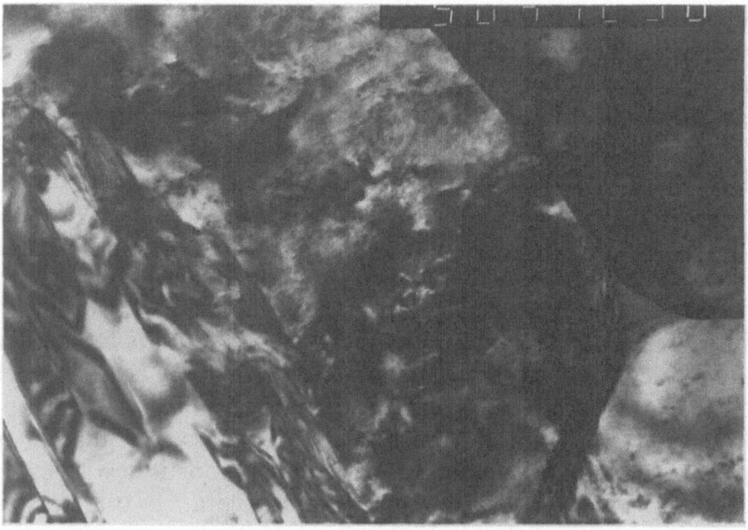

Fig. 4b. TEM of a Ce–TZP showing monoclinic twins and grain boundary glass phase at the grain triple points (magnification ×50 000).

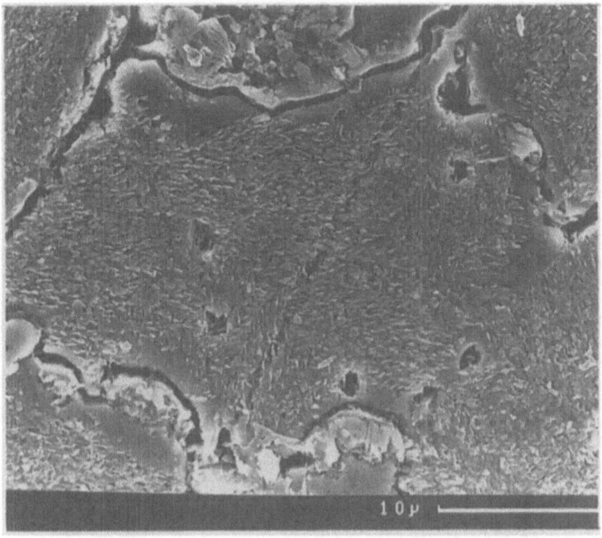

Fig. 4c. SEM of Mg–PSZ showing large cubic grains containing tetragonal precipitates with grain boundary impurity phase.

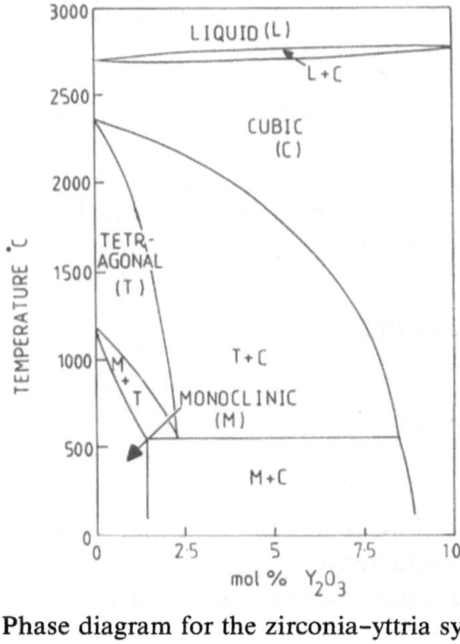

Fig. 5. Phase diagram for the zirconia–yttria system.

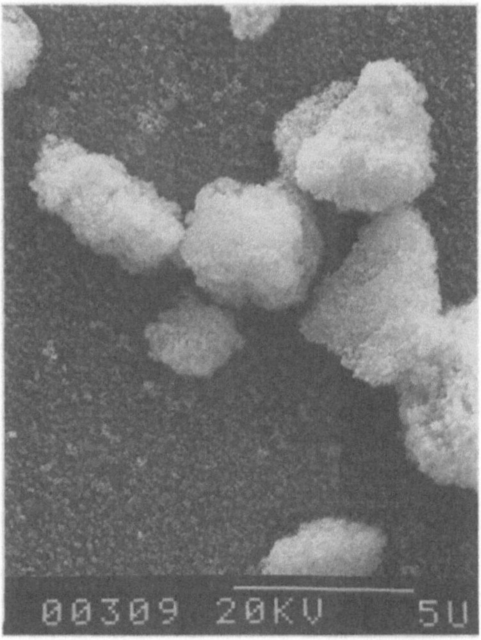

Fig. 6a. Chemically produced primary powder particles (3Y–TZP) in de-agglomerated form prior to spray drying, as shown in Fig. 6b.

4. PROPERTIES OF TZP

4.1 Mechanical and Physical Properties

The fundamental properties of TZPs which are of interest to the engineer or designer are:

high strength,
high fracture toughness,
high hardness,
wear resistance,
good frictional behaviour,
anti-static,
non-magnetic,
electrical insulation,
low thermal conductivity,
corrosion resistance in acids and alkalis,

Fig. 6b. Spray dried 3Y–TZP powder.

good surface finish, capability,
modulus of elasticity similar to steel,
coefficient of thermal expansion similar to iron.

Typical mechanical and physical properties are detailed in Table 1.

In common with all other engineering ceramics, the attainment of the above properties is largely dependent on both the starting powders and the fabrication techniques. All ceramic consolidation techniques have been applied to TZPs, with a typical process flow diagram detailed in Fig. 8. Flaw elimination at all process stages is crucial for not only high strength but also high reliability.[11] With critical flaws of the order of 20 μm, clean-room processing (Fig. 9) has been shown to significantly enhance both mean strengths and the distribution of strengths as measured by the Weibull modulus. Examples of critical defects which can occur in TZPs are shown in Figs 10 and 11. Both micrographs are taken from the fractured surfaces of three-point bend strength test bars.

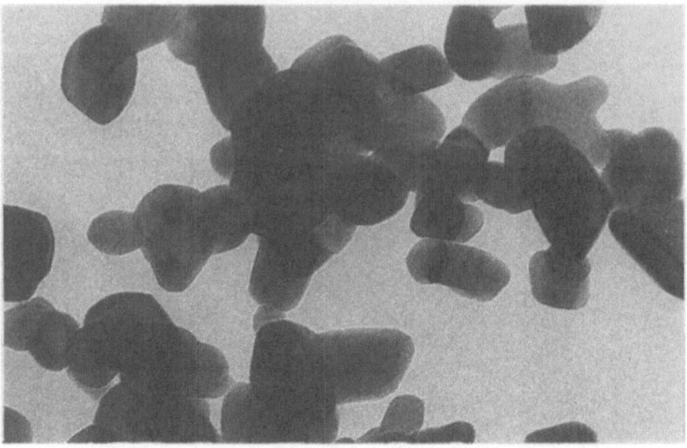

Fig. 7a. Uncoated, plasma synthesised Y–TZP powder (magnification ×200 000)

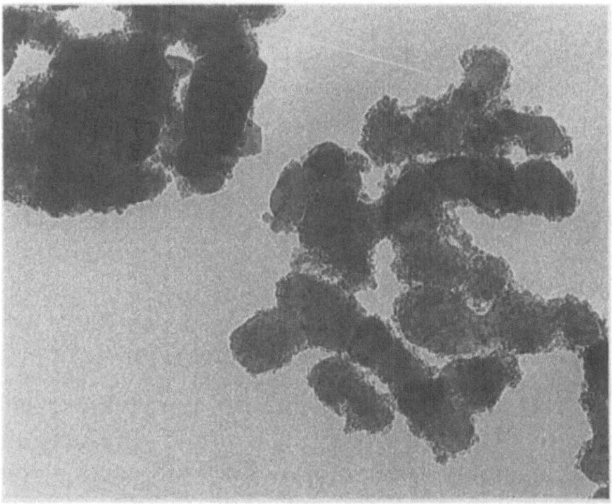

Fig. 7b. Plasma synthesised and yttria coated TZP powder (magnification ×200 000).

The defect shown in Fig. 10 can be attributed to 'unclean' processing and would result in a measured strength reduction of 20% when compared to an optimally processed TZP. The large cubic grains in Fig. 11 which represent the critical flaw in this specimen, demonstrate the effect of an inhomogeneous stabiliser distribution in the starting

TABLE 1
Comparative mechanical and physical properties for zirconia ceramic alloys.

	Y–TZP	Ce–TZP	ZTA	Mg–PSZ	3Y20A[a]
Density (Mg m^{-3})	6·05	6·15	4·15	5·75	5·51
Hardness (HV_{30})	1 350	900	1 600	1 020	1 470
Bend strength (MPa)	1 000	350	500	800	2 400
Compressive strength (MPa)	>2 000	—	—	1 850	—
Young's modulus (GPa)	205	215	380	205	260
Poisson's ratio	0·3	—	—	0·23	—
Fracture toughness (MPa m^{½})	9·5	15–20	4–5	9	6
Thermal expansion coefficient (MK^{-1})	10	8	8	10	9·4
Thermal conductivity (W m^{-1} K^{-1})	2	2	23	1·8	3
Surface finish (Ra/μm)	<0·02	—	—	<0·02	—

Y–TZP: yttria – tetragonal zirconia polycrystals.
Ce–TZP: ceria – tetragonal zirconia polycrystals.
ZTA: zirconia toughened alumina.
Mg–PSZ: magnesia – partially stabilised zirconia.
3Y20A: 3 mol % yttria zirconia 20 wt% alumina.

Following HIPing

Note: The values quoted are typical, but should only be used for guidance.
[a]Following HIPing.

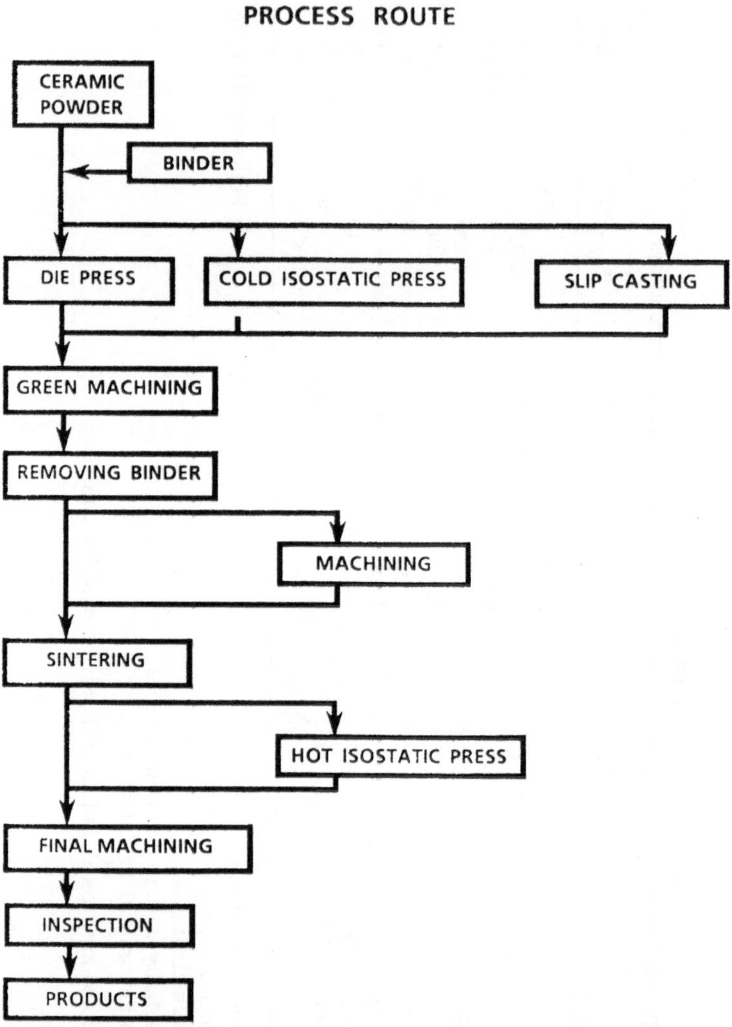

Fig. 8. Process flow diagram for the production of TZP ceramics.

powder. For TZPs processed under ideal conditions, mean strengths of 1150 MPa with a Weibull modulus of >30 can be achieved.

The combination of strength and toughness which can be achieved with TZP alloys is detailed in Fig. 12.[12] The characteristic peaks in individual alloy curves have been explained by Swain and Rose[13] as the transition between inherent flaw limited strength at low toughness, to

Fig. 9. Clean air powder processing facility.

transformation limited strength at higher values of toughness. For the engineer, who can remain oblivious to the academic explanations, the freedom of choice between high strength and toughness, or a combination of the two is all that matters.

The (2Y–TZP)20A composite ($2Y/Al_2O_3$ in Fig. 12) has the distinction of being reported to be the strongest polycrystalline ceramic ever tested. It also contradicts the view that TZP materials are only acceptable for low temperature applications as it displays a bend strength of 1000 MPa at 1000°C. The likely explanation for the increased strength, when compared to other TZPs is a progressive decrease in the flaw size responsible for fracture. This is due to the grain growth inhibition of the zirconia by the dispersed alumina phase during hot pressing. Thus the alumina addition may be affecting the critical flaw size rather than the fracture energy.

The ceria-TZP (Ce–PSZ in Fig. 12) is remarkable for its toughness value which is well into the regime of metallic materials. Using the common toughness measurement technique of Vickers indentation, with most ceramic materials, radial cracking at the corners of the diamond indent is visible. However with Ce–TZPs no cracking is observed and only a rumpling of the surface around the indent is visible.

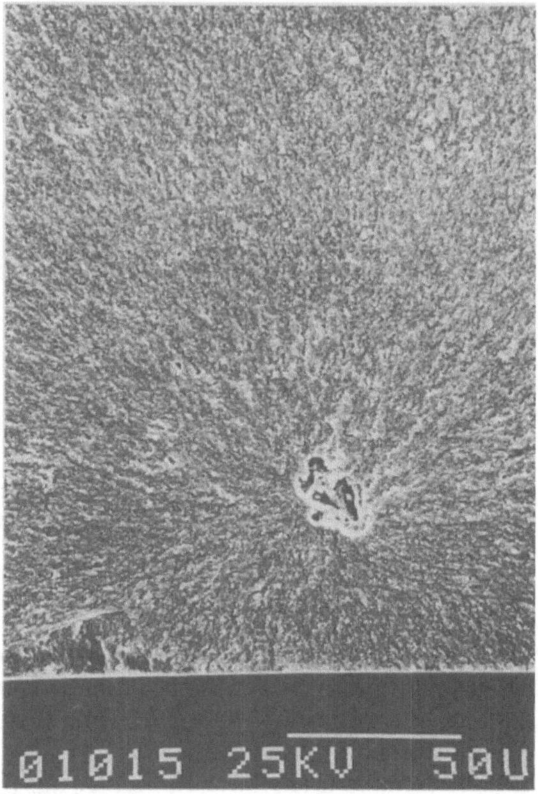

Fig. 10. Fracture surface of a TZP bend strength test bar illustrating a critical flaw due to unclean processing.

Interestingly for a ceramic structure prior to fracture, an optimum toughness Ce–TZP can display pseudo-plastic behaviour, with post-yield permanent strain. This damage tolerance introduces the previously unknown ceramic concept of 'graceful failure'. The high toughness also leads to a greater tolerance of processing defects, due to the increased size of the critical Griffith flaw in this material.

In common with many ceramic materials, hot isostatic pressing (HIPing) has been shown to increase substantially the measured strength values of Y–TZPs[14] by the elimination of internal defects. Such treatments can increase the strength of a 3Y–TZP by 50%, compared with normal pressureless sintered materials. Even though the measured

Fig. 11. Fracture surface of a TZP bend strength test bar illustrating a critical flaw due to large cubic grains.

density increase may only be minimal, i.e. 0·5%. HIPing in argon with graphite heaters also produces an interesting colour change in Y-TZP. The material changes colour from an ivory white to a silver grey or black. This colour change has been attributed to oxygen non-stoichiometry,[15] and carbothermal reduction effects.[16] In both of these studies the discoloration effect has been shown to have a detrimental effect on the high temperature strength and ageing behaviour of the Y-TZPs. Samples HIPed in an oxygen atmosphere with platinum/rhodium elements displayed higher HIPed strengths (20% higher than conventionally HIPed materials) and a greater strength retention when aged at 1000°C for 1200 h. Under these conditions the poor ageing

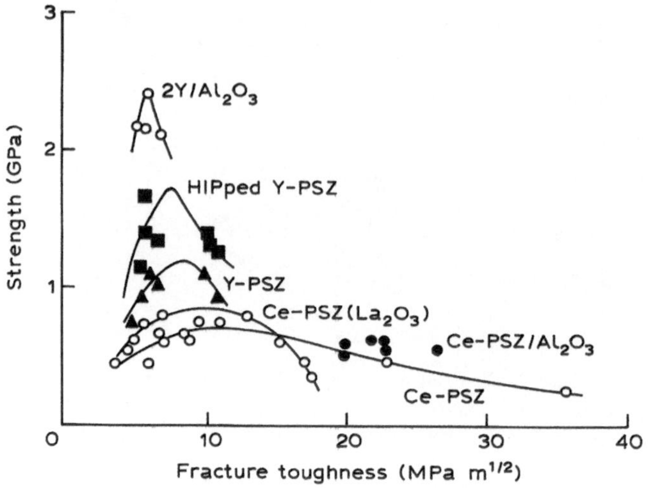

Fig. 12. Strength/toughness relationships in zirconia ceramics.

behaviour of Y-TZPs has been attributed to the presence of carbon. By ageing in air the carbon disappears and at the same time small pores are regenerated along the grain boundary, leading to a concomitant decrease in strength.

4.2 Ageing Behaviour

Soon after the development of Y-TZP, several workers[17-19] discovered that under dry and moist atmospheres in the temperature range 200–300 °C, certain forms of TZP suffered catastrophic degradation; the degradation taking the form of surface destabilisation of the tetragonal phase followed by a progressive inward deterioration, until certain grades of Y-TZP literally turn to dust. Although it is generally accepted that the removal of the elastic constraint at the grain boundary is the cause of the phenomenon, the mechanism for the degradation of properties with time and temperature is still the subject of discussion. The scheme proposed by Sato and Shimada[17] is analogous to the corrosion mechanism of vitreous silica by water, i.e. the debonding of the zirconium–oxygen bond by the presence of a proton donor. This mechanism is based on enhanced degradation by ageing the specimens in water, compared with dry air. The mechanism suggested by Lange *et al.*[18] is the destabilisation of the surface tetragonal zirconia grains, by a reaction of water vapour with yttrium. This suggestion is based on the experimental results observed by transmission electron microscopy examination (TEM) of thin foils, before and after ageing. The formation of clusters of small (20–50 nm) crystallites of alpha $Y(OH)_3$ on the surface of thin foils was confirmed by EDAX.

The susceptibility to degradation is also largely dependent on grain size and stabiliser content, or in other words transformability, and can now be avoided by modifying the powder morphology, particle size, sintering temperature and stabiliser distribution.[9] Other workers[20] have utilised the ageing behaviour of TZPs in air at 250°C to enhance the toughness of a 3Y–TZP. In this work the controlled transformation of surface layers from the tetragonal to monoclinic structure lead to the generation of compressive surface stresses and enhanced surface toughness (as measured by indentation).

4.3 Wear Behaviour

The wear behaviour of ceramics is often neglected by engineers, as they assume that the intrinsic hardness of ceramics will provide superior wear resistance. A more detailed analysis reveals that a static room temperature property such as hardness can not be used in isolation to predict wear behaviour. Instead it is necessary to consider the interaction of all material parameters (mechanical/physical properties and microstructure) with the operating environment.

When one considers the nature of the wearing interface for a TZP/ceramic or TZP/metal pair, in addition to environmental factors such as load and speed, one must also bear in mind the effect of transformation toughening on the surface of the zirconia component, as this is directly related to the fracture toughness and hence the resistance to indentation fracture. With TZP wire drawing dies it has been shown that high toughness can have a deleterious effect on performance[21] (Figs 13a, 13b). The wear process on the high toughness die being caused by surface transformation and grain removal. As illustrated diagrammatically in Fig. 14.

It is now understood that due to the low thermal conductivity of TZP, like-on-like combinations do not perform well at high loads and speeds, due to the contribution of frictional heating. Such surface heating can substantially modify room temperature properties such as hardness, and has been shown to facilitate intimate chemical mixing of TZP/ZTA sliding interfaces.[22] However if a TZP is mated against a thermally conductive counterface such as bearing steel,[23] the wear performance is several orders of magnitude better than silicon carbide, silicon nitride

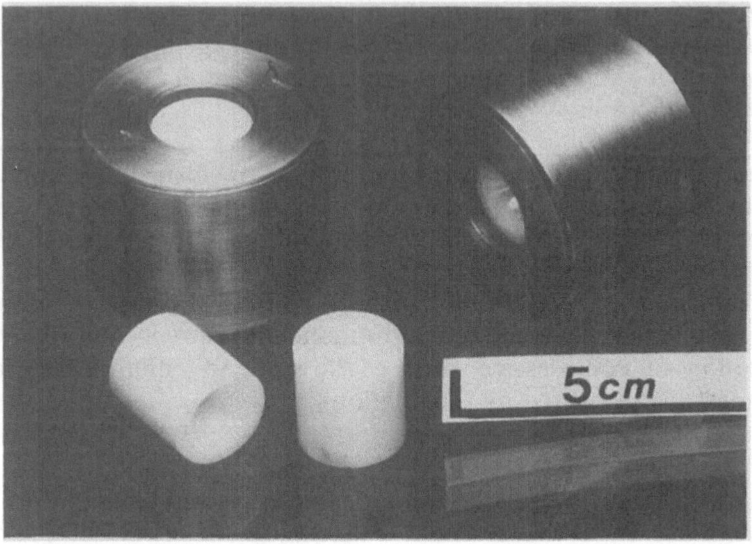

Fig. 13a. TZP wire drawing dies.

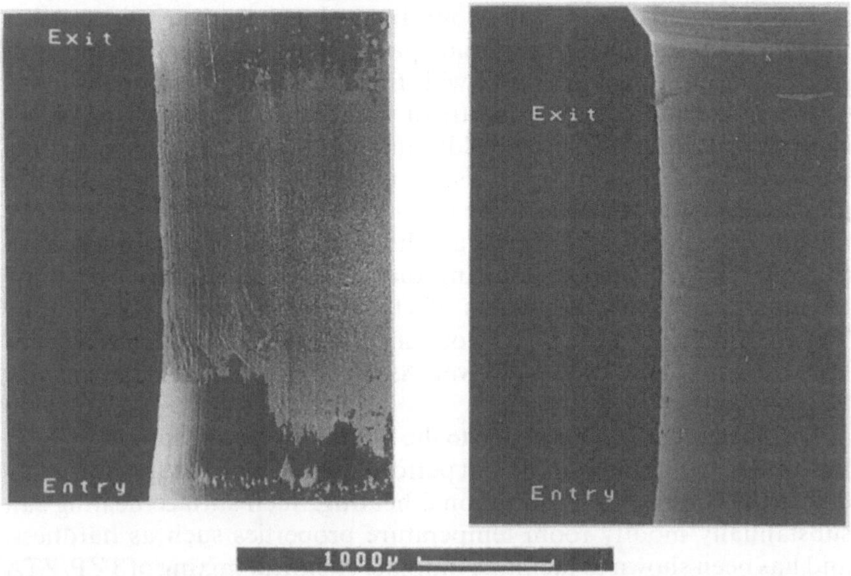

Fig. 13b. TZP wire drawing dies in section. SEM micrographs of the worn surfaces. The left hand micrograph shows high wear due to high transformability. The right hand micrograph shows the optimum wear behaviour of the less transformable TZP.

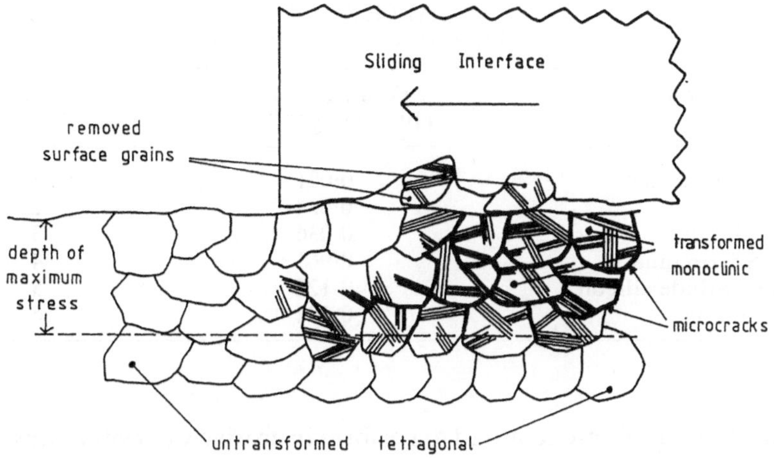

Fig. 14. Schematic illustrating the wear process in highly transformable TZPs.

and alumina sliding against the same counterface (see Table 2). In similar work, Becker *et al.*[24] have shown that unlike all other advanced ceramics, optimum toughness TZPs do not display the lateral crack subsurface fracture which causes surface spalling and accelerated abrasive wear. This conclusion implies that although a TZP displays lower indentation hardness than silicon carbide, due to the absence of lateral cracking the bulk wear of the TZP can be lower.

The high degree of surface finish which can be attained on TZPs also significantly improves both the initial and long term wear behaviour, as ceramics do not display the 'running in' behaviour or surface asperity rounding which is evident in metallic materials.

In addition to the sliding wear behaviour discussed above, in many

TABLE 2
The coefficient of friction and specific wear rate of a bearing steel ball on various ceramic discs

Material	Coefficient of friction	Specific wear rate ($mm^2\ N^{-1}$)
Silicon carbide	0·60	$1{\cdot}1 \times 10^{-9}$
Silicon nitride	0·83	$1{\cdot}4 \times 10^{-9}$
Alumina	0·52	$1{\cdot}6 \times 10^{-8}$
TZP zirconia	0·22	$6{\cdot}7 \times 10^{-11}$

The load was 10 N and sliding velocity was 17·5 cm s^{-1}.

TABLE 3
Comparative erosive wear performance

Material	Volume erosion (10^{-3} cm^3 g^{-1} (SiC))	Ranking
HP boron carbide	0·021	1
TZP zirconia	0·024	2
99·5% alumina	0·036	3
Mg–PSZ zirconia	0·061	4
Silicon carbide/silicon	0·125	5
93% alumina	0·101	6

applications TZPs are required to perform in erosive wear situations. In this wear regime the elevated fracture toughness/hardness relationship of TZPs becomes the dominant parameter in determining wear behaviour. It has been shown that Y–TZP has similar performance characteristics to much harder materials such as boron carbide and hot pressed silicon nitride.[25, 26] In the same work it has also been shown to be superior to magnesia partially stabilised zirconia (Mg–PSZ) which has a similar toughness, but a reduced hardness (see Table 3).

4.4 Superplastic Forming

The superplastic forming of TZPs represents the lateral thinker's approach to the relatively poor high temperature creep resistance of TZPs. Turning a detrimental mechanical property into an exciting and novel ceramic forming technique. The creep of TZPs at temperatures above 1100°C occurs readily due to the lack of transformation toughening and the small uniform grain size which aids plastic flow. The uniaxial tensile deformation behaviour of Y–TZP has been studied at temperatures up to 1500°C in ambient atmospheres by Wakai *et al.*[27] This work has shown that tensile test specimens can be superplastically deformed by 120% and still retain room temperature strengths of more than 950 MPa. In test samples which have been strained rapidly, extensive cavitation damage is visible, but at lower strain rates ($1{\cdot}1 \times 10^{-4}$ s^{-1}) cavitation effects were reduced. This work also highlighted that the strain rate was proportional to the inverse square of the grain size.

Clearly, although the principle of superplastic forming has been established, it will now take some considerable development by the component producer to develop this technology into a viable 'hot forging' production technique.

5. APPLICATIONS

For a period of perhaps 8 years from the initial discovery of transformation toughening, zirconia ceramics were viewed as a potential 'wonder' material for the future of engineering, science and technology. The levels of enthusiasm and 'hype' were at a similar level to the interest now being expressed in ceramic superconductors. Then, as now, much of this bullish behaviour was fuelled by wildly optimistic market surveys for the future potential of zirconia ceramics in a wide range of untested applications. Consequently, when the 'mega-markets' failed to materialise the academic and industrial enthusiasm for zirconia ceramics began to wane.

With the benefit of hindsight one can now highlight two fundamental reasons for the lack of immediate market penetration. First, the predicted large markets for zirconia (both Y–TZP and Mg–PSZ) were primarily in the field of engine ceramics, where the benefits of insulation and wear resistance were expected to revolutionise engine performance. These expectations foundered due to the immaturity of the material, the arduous nature of the application and the lack of tribological data. Second, in more general engineering components, zirconia ceramics saw relatively low utilisation due to the limited 'ceramic design' abilities of engineers, who, in addition to a limited knowledge of the properties of zirconia, almost inevitably favour direct substitution of the ceramic for its metal counterpart. This educational limitation is now being addressed by both government[28] and responsible suppliers.

Consequently, many of the applications of TZPs which are detailed below, have taken several months, if not years of considered design and material development to come to fruition. An indication of the range of shapes, sizes and components which can be produced in TZP is shown in Fig. 15.

5.1 Pumping Components

The excellent erosive wear and corrosion resistance of TZPs has led to their use in several pump components, typically parts subjected to high stress, such as shafts, couplings or thrust plates. NGK Insulators

Fig. 15. A selection of TZP components.

(Nagoya, Japan) have published details of a centrifugal pump which has all major parts in ceramic materials, with zirconia being used for the shaft, rotor cover and can. Typical application areas for such pumps are in sludge pumps and process pumps for the chemical industry.

Mechanical seal combinations of TZP and metallic counterfaces also show significant potential in abrasive slurry pumping (Fig. 16).

5.2 Engine Ceramics

TZPs have been considered for a range of wear resistance and insulation applications in internal combustion engines. Particularly in the valve train and cam followers[29,30] (Figs 17a, 17b). Other engine applications include the insulation of the combustion chamber, valve seats and as a coating on valve faces. To date none of these applications has been reported to be in full scale production.

5.3 Blade Edges

The combination of toughness, strength, hardness and a small grain size enables TZPs to display excellent edge retention, which in turn allows their use as blade edges. Originally developed in Japan as sushi knives, TZP cutting edges are now applied in blades for the paper

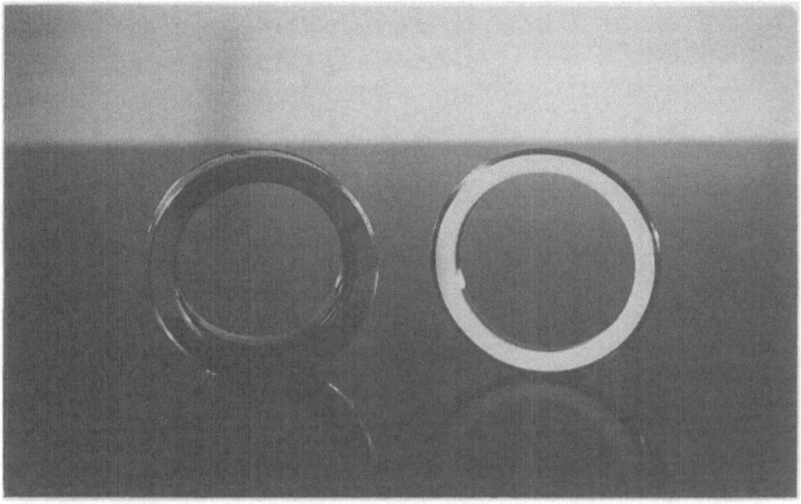

Fig. 16. TZP mechanical seal and metal counterface.

industry, medicine, hairdressing and the cutting of kevlar fibres. Although a TZP blade is no sharper than a conventional steel blade it will retain its edge up to 20 times longer in controlled tests (Fig. 18).

5.4 Consumer Components

In addition to the mechanical properties discussed earlier, TZP zirconia can also be aesthetically pleasing especially when polished. As a result, consumer items such as watch cases and straps have been produced. Sports items such as ice skate blades, football studs and golf putters have also been manufactured (Fig. 19). The consumers' acceptance of such items in Japan has also had the 'knock on effect' of a wider acceptance of these materials in more mundane areas throughout industry. Interestingly, with items such as the zirconia knife and scissors (Fig. 20), Japanese suppliers include a series of instructions on how to use the knife correctly, illustrating the point that with many of these applications the cultural preconceptions of the consumer must be addressed if successful commercialisation is to be achieved.

5.5 General Engineering

As the 'ceramic awareness' of industry increases, the number of critical components replaced by TZPs is also increasing. As individual

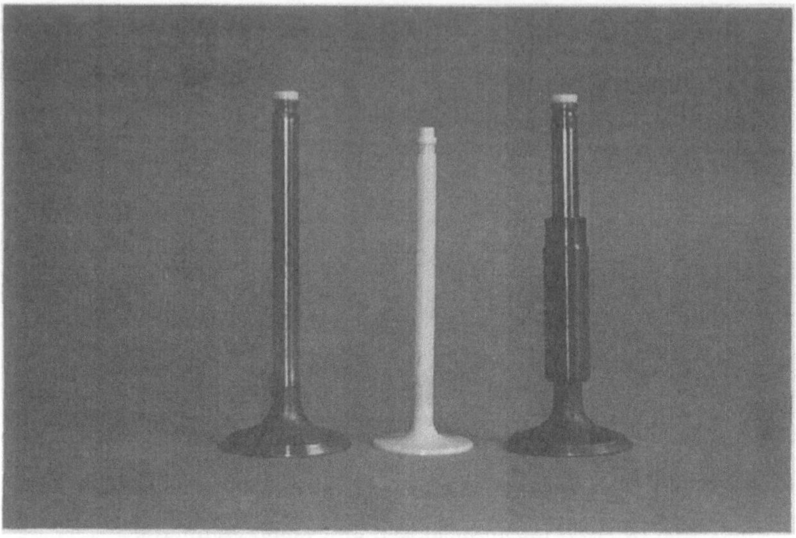

Fig. 17a. TZP engine ceramics, valve, valve guide and tip.

Fig. 17b. TZP rocker bushes.

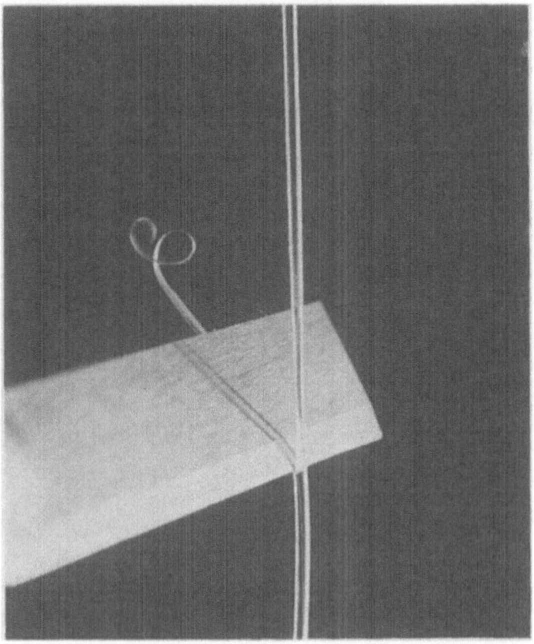

Fig. 18. TZP fibre cutting blade.

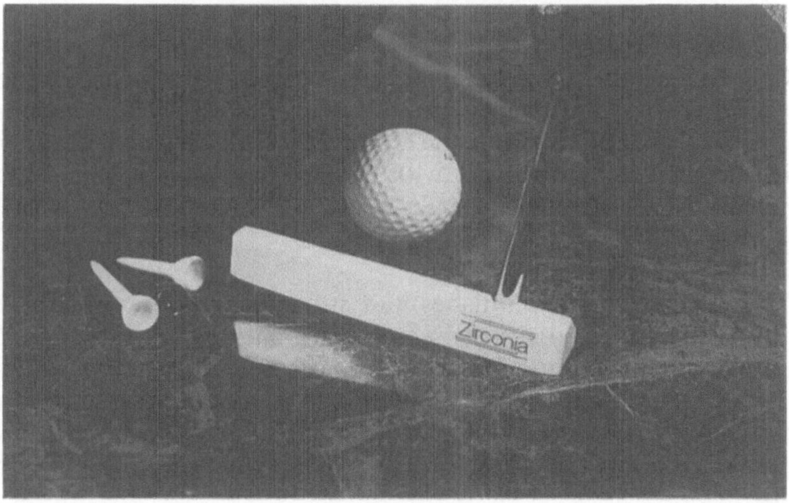

Fig. 19. TZP golf putter.

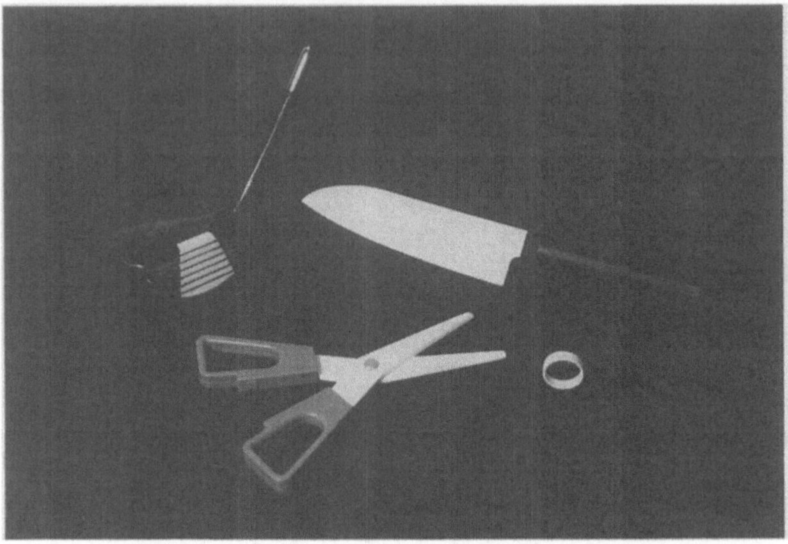

Fig. 20. TZP scissors, knife, golf club insert and ring.

applications they do not represent big volume markets but their cumulative effect is substantial. Typical applications include the wear resistant parts shown in Figs 21a–21e.[21]

5.6 Grinding Media

Grinding media currently represent the largest tonnage requirement for TZPs with a requirement of around 20 tonnes per year. The obvious benefits in this application are the high density of TZP which gives good milling efficiencies, coupled with wear resistance which reduces contamination.

5.6 Telecommunications

The excellent surface finishes and strengths which are attainable with TZPs has led to their use in optical fibre technologies where they are used as split coupling devices for optical fibres. In parallel with the injection moulding technology which has been developed to produce such components, diamond finishing techniques have also been improved to allow tolerances of 1 μm to be achieved on a repeatable basis.

6. CONCLUSIONS

The science and technology of zirconia ceramics, and in particular tetragonal zirconia polycrystals is continuing to evolve at a considerable pace. The material is still in a dynamic stage of development with new property and processing results being reported virtually every month. These material and property improvements, coupled with a greater ceramic understanding in the engineering community, are leading to increasing applications in many industrial sectors.

Many of the problems facing the zirconia producer are common to all other advanced ceramics, namely, the search for increased reliability, economic fabrication techniques and effective non-destructive testing, together with a lack of agreed standards. However, the specific zirconia issues concern their relatively poor high temperature strength and creep resistance (except the alumina–zirconia composite detailed in Section 4.1). In most cases this limits the structural applications of TZP's to below 1000 °C, due to the thermodynamic limitations of the transformation toughening mechanism above 800 °C. Claussen[31] has outlined how these deficiencies could be overcome by microstructural engineering strategies such as fibre reinforcement and special grain boundary design. The incorporation of 30 vol. % of silicon carbide whiskers has been shown to double the strength of Y–TZP at 1000 °C. Unfortunately, the fabrication of such whisker-reinforced composites generally requires hot-pressing which currently imposes shape and size limitations on the sintered part.

As a consequence of the high thermal expansion and low thermal conductivity of TZPs, the combination of these two properties leads to a relatively poor thermal down shock resistance. Therefore future work must address this issue by considering how thermally conductive agents can be incorporated into the TZP matrix without compromising the measured properties.

The future success of zirconia ceramics is also largely dependent on the responsible attitude of suppliers towards end users. The onus is on the component supplier to suggest the use of a zirconia ceramic only in areas where it is likely to work. If the environmental conditions fall outside the capabilities of zirconia it is imperative that this is recognised and the end user is directed towards more suitable materials. It is only by taking this approach that long term progress will be achieved at the expense of short term sales. Far too often in the recent past zirconia has

Fig. 21a. TZP ball mill, 500 ml capacity.

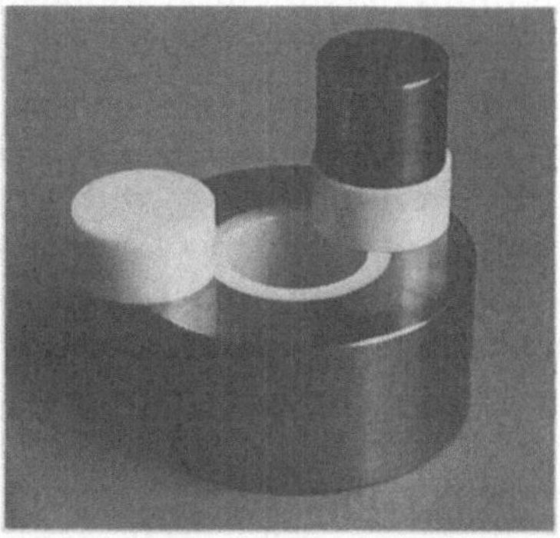

Fig. 21b. TZP die pressing toolset for powder metallurgy.

Fig. 21c. TZP spray nozzle and needle.

Fig. 21d. TZP pin in a pin type mixer.

Fig. 21e. TZP coil springs.

been applied in areas to which it was not suited. Subsequently the end user has held a negative opinion towards the material and is not inclined to risk his reputation with the material in future.

Hopefully, within the confines of this relatively selective review it will have become apparent that significant progress has been made with zirconia ceramics throughout the 1980s. They now represent an important class of 'enabling materials' for both current and future technologies, although considerable effort is still required by the material scientist, the marketeer and the end user before the title of 'ceramic steel' can be applied with absolute justification.

ACKNOWLEDGEMENTS

The authors would like to thank the following people for their help and assistance in the production of this review: Drs R. Stevens, G. P. Dransfield, W. M. Rainforth and Mr T. M. Allen.

REFERENCES

1. Garvie, R. C., Hannik, R. H. J. and Pascoe, R. T., Ceramic steel?, *Nature (London),* **258** (1975) 5537.
2. Stevens, R., *An introduction to zirconia*, 2nd edn, Magnesium Elecktron Publication No. 113, Magnesium Elecktron Inc., New Jersey, 1986.
3. Heuer, A. A. and Hobbs, L. W., *Advances in ceramics, Vol. 3, Science and technology of zirconia*, The American Ceramic Society, Ohio, 1981.
4. Claussen, N., Rühle, M. and Heuer, A. A., *Advances in ceramics, Vol. 12, Science and technology of zirconia II*, The American Ceramic Society, Ohio, 1983.
5. Somiya, S., Yamamoto, N. and Hanagida, H., *Advances in Ceramics, Vol. 24, Science and technology of zirconia III*, The American Ceramic Society, Ohio, 1988.
6. Lange, F. F., Transformation toughening: Parts 1–4, *J. Mater. Sci.,* **17** (1982) 225–54.
7. Tosoh Manufacturing Co., Tokyo, Technical Bulletin TZ3Y Ceramics.
8. Blackburn, S., Kerridge, C. R. and Senhenn, P. G., *Toughened zirconia ceramics from electro-refined PSZ powders*, Unitec Ceramics, Technical Publication, 1987, Stafford, UK.
9. Dransfield, G. P., Fothergill, K. A. and Egerton T. A., The use of plasma synthesis and pigment coating technology to produce an yttria stabilised zirconia having superior properties. Presented at the European Ceramic Society Conference, Maastricht, 1989.
10. Rhone-Poulenc Ltd, Technical Literature.

11. Allen, T. M., Birkby, I. and Stevens, R., 'Nature and effect of defects introduced during the fabrication of zirconia engineering ceramics, *Powder Metallurgy,* **31** (1) (1988) 23.
12. Tsukuma, K., Ueda, K. and Shiomi, M., Mechanical properties of isostatically hot-pressed zirconia (yttria)/alumina composites. Presented at the 38th Annual Pacific Coast Regional Meeting of American Ceramic Society, Irvine California, 1985.
13. Swain, M. V. and Rose, L. R. F., Strength limitations of transformation toughened zirconia alloys, *J. Am. Ceram. Soc.,* **69** (7) (1986) 511.
14. Tsukuma, K., Kubota, Y. and Tsukidate, T., Thermal and mechanical properties of yttria-stabilised tetragonal zirconia polycrystals, *Advances in ceramics.,* Vol. 12, The American Ceramic Society, Ohio, 1983.
15. Swab, J. J., Katz, R. N. and Starita, C. J., Effects of oxygen nonstoichemetry on the high temperature performance of an yttria tetragonal zirconia polycrystal material, *Advanced Ceramic Materials, Vol. 3,* No. 3, American Ceramic Society, Ohio, 1988.
16. Manabe, Y., Fujikawa, T., Ueda, M. and Inoue, Y., Effect of oxygen HIP for oxide ceramics. Presented at the 1st European Ceramic Society Conference, held at Maastricht, Netherlands, June 18–23, 1989.
17. Sato, T. and Shimada, M., Transformation of yttria doped tetragonal zirconia polycrystals by annealing in water, *J. Am. Ceram. Soc.,* **68** (1985) 356.
18. Lange, F. F., Dunlop, G. L. and Davis, B. I., Degradation during aging of transformation-toughened ZrO_2-Y_2O_3 materials at 250 °C. *J. Am. Ceram. Soc.,* **69** (1986) 237.
19. Masaki, T., Mechanical properties of yttria partially stabilised zirconia after aging at low temperatures, *Int. J. High Tech. Ceram.,* **2** (1986).
20. Wang, J. and Stevens, R., Surface toughening of TZP ceramics by low temperature ageing, *Ceram. Int.,* **15** (1989) 15.
21. Birkby, I., Harrison, P. and Stevens, R., The effect of surface transformations on the wear behaviour of zirconia TZP ceramics, *J. Europ. Ceram. Soc.,* **5** (1989).
22. Rainforth, W. M., Stevens, R. and Nutting, J., Observations on the sliding wear of transformation toughened ceramics. Presented at the 1st European Ceramic Society Conference held at Maastricht, Netherlands, June 18–23, 1989.
23. Iwasa, M. and Kinoshita, M., Friction and wear of bearing steel sliding on ceramics measured by ball on disc method, *J. Ceram. Soc. Japan,* **95** (9) (1987) 899.
24. Becker, P. C., Libsch, T. A. and Rhee, S. K., Wear mechanisms of toughened zirconias, Ceramic Engineering and Science Procs, Conf. on Automotive Materials, American Ceramic Society, 1985.
25. Kingon, A. J., Erosive wear of some zirconia based ceramics, Procs 6th Cimtec Conf. on High Tech Ceramics, Milan, Italy, 1986.
26. Wada, S. and Watanabe, N., Erosion rate of zirconia ceramics, *J. Ceram. Soc. Japan,* **96** (5) (1988) 599.
27. Wakai, F., Sakaguchi, S. and Yosoo, M., Superplasticity of TZP's, *Advanced Ceramic Materials,* **1** (3) (1986) 259.

28. *Materials Matter*, Dept of Trade and Industry Technical Publication, HMSO, 1989.
29. The CARE Consortium, Department of Trade and Industry Technical Publication, HMSO, 1988.
30. Fingerle, D., Gundel, W. and Olapinski, H., Friction and wear reduction by ceramic components, *Proc. 2nd Int. Conf. on Ceramic Materials and Components for Engines*, Lubeck-Travemunde, 1986.
31. Claussen, N., Strengthening strategies for zirconia toughened ceramics (ZTC) at high temperatures, *Mater. Sci. Engng*, **71** (23) (1985).

Index